T0303916

GENETICS, GENOMICS AND BREEDING OF PEPPERS AND EGGPLANTS

Genetics, Genomics and Breeding of Crop Plants

Series Editor
Chittaranjan Kole
Department of Genetics and Biochemistry
Clemson University
Clemson, SC
USA

Books in this Series:

Published or in Press:

- Jinguo Hu, Gerald Seiler & Chittaranjan Kole: *Sunflower*
- Kristin D. Bilyeu, Milind B. Ratnaparkhe & Chittaranjan Kole: *Soybean*
- Robert Henry & Chittaranjan Kole: *Sugarcane*
- Kevin Folta & Chittaranjan Kole: *Berries*
- Jan Sadowsky & Chittaranjan Kole: *Vegetable Brassicas*
- James M. Bradeen & Chittaranjan Kole: *Potato*
- C.P. Joshi, Stephen DiFazio & Chittaranjan Kole: *Poplar*
- Anne-Françoise Adam-Blondon, José M. Martínez-Zapater & Chittaranjan Kole: *Grapes*
- Christophe Plomion, Jean Bousquet & Chittaranjan Kole: *Conifers*
- Dave Edwards, Jacqueline Batley, Isobel Parkin & Chittaranjan Kole: *Oilseed Brassicas*
- Marcelino Pérez de la Vega, Ana María Torres, José Ignacio Cubero & Chittaranjan Kole: *Cool Season Grain Legumes*
- Yi-Hong Wang, Tusar Kanti Behera & Chittaranjan Kole: *Cucurbits*
- Albert G. Abbott & Chittaranjan Kole: *Stone Fruits*
- Barbara E. Liedl, Joanne A. Labate, John R. Stommel, Ann Slade & Chittaranjan Kole: *Tomato*
- Byoung-Cheorl Kang & Chittaranjan Kole: *Peppers and Eggplants*

GENETICS, GENOMICS AND BREEDING OF PEPPERS AND EGGPLANTS

Editors

Byoung-Cheorl Kang
Dept. of Plant Science
College of Agriculture & Life Sciences
Seoul National University
Seoul
Korea

Chittaranjan Kole
Department of Genetics and Biochemistry
Clemson University
Clemson, SC
USA

CRC Press
Taylor & Francis Group
Boca Raton   London   New York

CRC Press is an imprint of the
Taylor & Francis Group, an **informa** business

A SCIENCE PUBLISHERS BOOK

CRC Press
Taylor & Francis Group
6000 Broken Sound Parkway NW, Suite 300
Boca Raton, FL 33487-2742

© 2013 Copyright reserved
CRC Press is an imprint of Taylor & Francis Group, an Informa business

No claim to original U.S. Government works

Printed in the United States of America on acid-free paper

International Standard Book Number: 978-1-4665-7745-9 (Hardback)

Library of Congress Cataloging-in-Publication Data

Genetics, genomics and breeding of peppers and eggplants / editors:
Byoung-Cheorl Kang, Chittaranjan Kole.
        p. cm. -- (Genetics, genomics and breeding of crop plants)
   Includes bibliographical references and index.
   ISBN 978-1-4665-7745-9 (hardcover : alk. paper)   1.
Peppers--Genetics. 2. Peppers--Genome mapping. 3. Peppers--
Breeding.
4. Eggplant--Genetics. 5. Eggplant--Genome mapping. 6.
Eggplant--Breeding. I. Kang, Byoung-Cheorl. II. Kole,
Chittaranjan.
III. Series: Genetics, genomics and breeding of crop plants.
   SB351.P4G46 2013
   583'.25--dc23

                                                      2012036520

Visit the Taylor & Francis Web site at
http://www.taylorandfrancis.com

CRC Press Web site at
http://www.crcpress.com

Science Publishers Web site at
http://www.scipub.net

Preface to the Series

Genetics, genomics and breeding has emerged as three overlapping and complimentary disciplines for comprehensive and fine-scale analysis of plant genomes and their precise and rapid improvement. While genetics and plant breeding have contributed enormously towards several new concepts and strategies for elucidation of plant genes and genomes as well as development of a huge number of crop varieties with desirable traits, genomics has depicted the chemical nature of genes, gene products and genomes and also provided additional resources for crop improvement.

In today's world, teaching, research, funding, regulation and utilization of plant genetics, genomics and breeding essentially require thorough understanding of their components including classical, biochemical, cytological and molecular genetics; and traditional, molecular, transgenic and genomics-assisted breeding. There are several book volumes and reviews available that cover individually or in combination of a few of these components for the major plants or plant groups; and also on the concepts and strategies for these individual components with examples drawn mainly from the major plants. Therefore, we planned to fill an existing gap with individual book volumes dedicated to the leading crop and model plants with comprehensive deliberations on all the classical, advanced and modern concepts of depiction and improvement of genomes. The success stories and limitations in the different plant species, crop or model, must vary; however, we have tried to include a more or less general outline of the contents of the chapters of the volumes to maintain uniformity as far as possible.

Often genetics, genomics and plant breeding and particularly their complimentary and supplementary disciplines are studied and practiced by people who do not have, and reasonably so, the basic understanding of biology of the plants for which they are contributing. A general description of the plants and their botany would surely instill more interest among them on the plant species they are working for and therefore we presented lucid details on the economic and/or academic importance of the plant(s); historical information on geographical origin and distribution; botanical origin and evolution; available germplasms and gene pools, and genetic and cytogenetic stocks as genetic, genomic and breeding resources; and basic information on taxonomy, habit, habitat, morphology, karyotype, ploidy level and genome size, etc.

Classical genetics and traditional breeding have contributed enormously even by employing the phenotype-to-genotype approach. We included detailed descriptions on these classical efforts such as genetic mapping using morphological, cytological and isozyme markers; and achievements of conventional breeding for desirable and against undesirable traits. Employment of the *in vitro* culture techniques such as micro- and megaspore culture, and somatic mutation and hybridization, has also been enumerated. In addition, an assessment of the achievements and limitations of the basic genetics and conventional breeding efforts has been presented.

It is a hard truth that in many instances we depend too much on a few advanced technologies, we are trained in, for creating and using novel or alien genes but forget the infinite wealth of desirable genes in the indigenous cultivars and wild allied species besides the available germplasms in national and international institutes or centers. Exploring as broad as possible natural genetic diversity not only provides information on availability of target donor genes but also on genetically divergent genotypes, botanical varieties, subspecies, species and even genera to be used as potential parents in crosses to realize optimum genetic polymorphism required for mapping and breeding. Genetic divergence has been evaluated using the available tools at a particular point of time. We included discussions on phenotype-based strategies employing morphological markers, genotype-based strategies employing molecular markers; the statistical procedures utilized; their utilities for evaluation of genetic divergence among genotypes, local landraces, species and genera; and also on the effects of breeding pedigrees and geographical locations on the degree of genetic diversity.

Association mapping using molecular markers is a recent strategy to utilize the natural genetic variability to detect marker-trait association and to validate the genomic locations of genes, particularly those controlling the quantitative traits. Association mapping has been employed effectively in genetic studies in human and other animal models and those have inspired the plant scientists to take advantage of this tool. We included examples of its use and implication in some of the volumes that devote to the plants for which this technique has been successfully employed for assessment of the degree of linkage disequilibrium related to a particular gene or genome, and for germplasm enhancement.

Genetic linkage mapping using molecular markers have been discussed in many books, reviews and book series. However, in this series, genetic mapping has been discussed at length with more elaborations and examples on diverse markers including the anonymous type 2 markers such as RFLPs, RAPDs, AFLPs, etc. and the gene-specific type 1 markers such as EST-SSRs, SNPs, etc.; various mapping populations including F_2, backcross, recombinant inbred, doubled haploid, near-isogenic and pseudotestcross; computer software including MapMaker, JoinMap, etc. used; and different types of genetic maps including preliminary, high-resolution, high-density, saturated, reference, consensus and integrated developed so far.

Mapping of simply inherited traits and quantitative traits controlled by oligogenes and polygenes, respectively has been deliberated in the earlier literature crop-wise or crop group-wise. However, more detailed information on mapping or tagging oligogenes by linkage mapping or bulked segregant analysis, mapping polygenes by QTL analysis, and different computer software employed such as MapMaker, JoinMap, QTL Cartographer, Map Manager, etc. for these purposes have been discussed at more depth in the present volumes.

The strategies and achievements of marker-assisted or molecular breeding have been discussed in a few books and reviews earlier. However, those mostly deliberated on the general aspects with examples drawn mainly from major plants. In this series, we included comprehensive descriptions on the use of molecular markers for germplasm characterization, detection and maintenance of distinctiveness, uniformity and stability of genotypes, introgression and pyramiding of genes. We have also included elucidations on the strategies and achievements of transgenic breeding for developing genotypes particularly with resistance to herbicide, biotic and abiotic stresses; for biofuel production, biopharming, phytoremediation; and also for producing resources for functional genomics.

A number of desirable genes and QTLs have been cloned in plants since 1992 and 2000, respectively using different strategies, mainly positional cloning and transposon tagging. We included enumeration of these and other strategies for isolation of genes and QTLs, testing of their expression and their effective utilization in the relevant volumes.

Physical maps and integrated physical-genetic maps are now available in most of the leading crop and model plants owing mainly to the BAC, YAC, EST and cDNA libraries. Similar libraries and other required genomic resources have also been developed for the remaining crops. We have devoted a section on the library development and sequencing of these resources; detection, validation and utilization of gene-based molecular markers; and impact of new generation sequencing technologies on structural genomics.

As mentioned earlier, whole genome sequencing has been completed in one model plant (Arabidopsis) and seven economic plants (rice, poplar, peach, papaya, grapes, soybean and sorghum) and is progressing in an array of model and economic plants. Advent of massively parallel DNA sequencing using 454-pyrosequencing, Solexa Genome Analyzer, SOLiD system, Heliscope and SMRT have facilitated whole genome sequencing in many other plants more rapidly, cheaply and precisely. We have included extensive coverage on the level (national or international) of collaboration and the strategies and status of whole genome sequencing in plants for which sequencing efforts have been completed or are progressing currently. We have also included critical assessment of the impact of these genome initiatives in the respective volumes.

Comparative genome mapping based on molecular markers and map positions of genes and QTLs practiced during the last two decades of the last century provided answers to many basic questions related to evolution, origin and phylogenetic relationship of close plant taxa. Enrichment of genomic resources has reinforced the study of genome homology and synteny of genes among plants not only in the same family but also of taxonomically distant families. Comparative genomics is not only delivering answers to the questions of academic interest but also providing many candidate genes for plant genetic improvement.

The 'central dogma' enunciated in 1958 provided a simple picture of gene function—gene to mRNA to transcripts to proteins (enzymes) to metabolites. The enormous amount of information generated on characterization of transcripts, proteins and metabolites now have led to the emergence of individual disciplines including functional genomics, transcriptomics, proteomics and metabolomics. Although all of them ultimately strengthen the analysis and improvement of a genome, they deserve individual deliberations for each plant species. For example, microarrays, SAGE, MPSS for transcriptome analysis; and 2D gel electrophoresis, MALDI, NMR, MS for proteomics and metabolomics studies require elaboration. Besides transcriptome, proteome or metabolome QTL mapping and application of transcriptomics, proteomics and metabolomics in genomics-assisted breeding are frontier fields now. We included discussions on them in the relevant volumes.

The databases for storage, search and utilization on the genomes, genes, gene products and their sequences are growing enormously in each second and they require robust bioinformatics tools plant-wise and purpose-wise. We included a section on databases on the gene and genomes, gene expression, comparative genomes, molecular marker and genetic maps, protein and metabolomes, and their integration.

Notwithstanding the progress made so far, each crop or model plant species requires more pragmatic retrospect. For the model plants we need to answer how much they have been utilized to answer the basic questions of genetics and genomics as compared to other wild and domesticated species. For the economic plants we need to answer as to whether they have been genetically tailored perfectly for expanded geographical regions and current requirements for green fuel, plant-based bioproducts and for improvements of ecology and environment. These futuristic explanations have been addressed finally in the volumes.

We are aware of exclusions of some plants for which we have comprehensive compilations on genetics, genomics and breeding in hard copy or digital format and also some other plants which will have enough achievements to claim for individual book volume only in distant future. However, we feel satisfied that we could present comprehensive deliberations on genetics, genomics and breeding of 30 model and economic plants, and their groups in a few cases, in this series. I personally feel also happy that I could work with many internationally celebrated scientists who edited the book volumes on the leading plants and plant groups and included chapters authored by many scientists reputed globally for their contributions on the concerned plant or plant group.

We paid serious attention to reviewing, revising and updating of the manuscripts of all the chapters of this book series, but some technical and formatting mistakes will remain for sure. As the series editor, I take complete responsibility for all these mistakes and will look forward to the readers for corrections of these mistakes and also for their suggestions for further improvement of the volumes and the series so that future editions can serve better the purposes of the students, scientists, industries, and the society of this and future generations.

Science publishers, Inc. has been serving the requirements of science and society for a long time with publications of books devoted to advanced concepts, strategies, tools, methodologies and achievements of various science disciplines. Myself as the editor and also on behalf of the volume editors, chapter authors and the ultimate beneficiaries of the volumes take this opportunity to acknowledge the publisher for presenting these books that could be useful for teaching, research and extension of genetics, genomics and breeding.

Chittaranjan Kole

Preface to the Volume

Peppers and eggplants are two leading vegetable crops produced and consumed worldwide. The consumption patterns of peppers are very diverse over cultures and regions. Peppers are the sole source for pungency and contain various functional nutrients such as vitamins and biological compounds. To facilitate the breeding for agronomical traits such as diseases resistance and quality traits, diverse molecular genetic studies have been performed. Although these studies have been hampered by disadvantageous characteristics of peppers such as large genome size and low efficiency of regeneration and transformation, recent achievements on pepper genome sequencing, trait-linked marker development enabled the cloning of genes involved in useful traits.

Solanum melongena is an old world species complex that includes weedy and wild relatives as well as primitive cultivars and landraces. As with peppers, the main breeding objectives are yield increase, disease resistance, and fruit quality using parthenocarpy and secondary metabolites. The close relationship between eggplant, tomato, and pepper has facilitated this work as well as made the Solanaceae a model for comparative genomics. In this book, the overall information on agronomical and evolutionary characteristics of peppers and eggplants and results of molecular genetic studies and genome structure are described.

Byoung-Cheorl Kang
Dept. of Plant Science
College of Agriculture & Life Sciences
Seoul National University
Seoul
Korea

Chittaranjan Kole
Department of Genetics and Biochemistry
Clemson University
Clemson, SC
USA

Contents

List of Contributors

Amy Frary
Mount Holyoke College, South Hadley, Massachusetts, USA.
Email: *afrary@mtholyoke.edu*

Byoung-Cheorl Kang
Department of Plant Science, Seoul National University, Seoul 151-921, Republic of Korea.
Email: *bk54@snu.ac.kr*

Cristina Fernández Otero
Centro de Investigaciones Agrarias de Mabegondo, A Coruña, Spain.
Email: *cristinafernandez@ciam.es*

Davis Cheng
Department of Biology, California State University Fresno, Fresno CA 93740 USA.
Email: *dcheng@csufresno.edu*

Doil Choi
Department of Plant Science, Seoul National University, Seoul 151-921, Republic of Korea.
Email: *doil@snu.ac.kr*

Hee-Bum Yang
Department of Plant Science, Seoul National University, Seoul 151-921, Republic of Korea.
Email: *yhbk0130@snu.ac.kr*

Hee-Jin Jeong
Department of Plant Science, Seoul National University, Seoul 151-921, Republic of Korea.
Email: *chungo30@snu.ac.kr*

Ilan Paran
Institute of Plant Sciences, The Volcani Center, Bet Dagan, Israel.
Email: *iparan@volcani.agri.gov.il*

James P. Prince
Department of Biology, California State University Fresno, Fresno CA 93740, USA.
Email: *jamespr@csufresno.edu*

Lindsay E. Wyatt
Department of Plant Breeding and Genetics, Cornell University, Ithaca, NY 14853.
Email: *lew67@cornell.edu*

Michael Mazourek
Department of Plant Breeding and Genetics, Cornell University, Ithaca, NY 14853.
Email: *mm284@cornell.edu*

MinKyu Park
Department of Plant Science, Seoul National University, Seoul 151-921, Republic of Korea.
Email: *minkju@hanmail.net*

Sami Doganlar
Izmir Institute of Technology, Izmir, Turkey.
Email: *samidoganlar@iyte.edu.tr*

Wing Yee Liu
School of Biological Sciences, The University of Hong Kong, 5N01, Kadoorie Biological Sciences Building, Pokfulam Road, Hong Kong.
Email: *winnielh@hku.hk*

Won-Hee Kang
Department of Plant Science, Seoul National University, Seoul 151-921, Republic of Korea.
Email: *wonhuijjang@hanmail.net*

Young Deuk Jo
Department of Plant Science, Seoul National University, Seoul 151-921, Republic of Korea.
Email: *cho1414@snu.ac.kr*

Abbreviations

AFLP	Amplified fragment length polymorphism
AT3	Acyltranferase 3
AVRDC	Asian Vegetable Research and Development Center
BAC	Bacterial artificial chromosome
BAHD	Benzylalcohol acetyltransferase (BEAT), anthocyanin-O-hydroxy-cinnamoyltransferase (AHCT), anthranilate N-hydroxycinnamoyl/benzoyltransferase (HCBT), and deacetylvindoline 4-O-acetyltransferase (DAT)
BAP	Benzyladenine
bar	Bialaphos resistance (gene)
BC	Backcross
BLA	Bulked line analysis
BSA	Bulk segregant analysis
CaMV	*Cauliflower mosaic virus*
CAPS	Cleaved amplified polymorphic sequences
CaSGR	*Capsicum* stay-green
CCS	Capsanthin capsorubin synthase
CENL	Capsicum and Eggplant Newsletter
CGMS	Cytoplasmic-genic male sterility
ChiVMV	*Chilli veinal mottle virus*
cl	Chlorophyll retainer
cM	Centi-Morgan
CMS	Cytoplasmic male sterility
CMV	*Cucumber mosaic virus*
COS	Conserved ortholog set
CP	Coat protein
cry	Cryptochrome (gene)
CS	Capsaicin synthase
dCAPS	Derived cleaved amplified polymorphic sequences
DH	Doubled haploid
dhfr	Dihydrofolate reductase
DUS	Distinctiveness, uniformity and stability
EB	Ethidium bromide

ECPGR	European Cooperative Programme for Plant Genetic Resources, Netherlands
EGGNET	Eggplant Genetic Resources Network
eIF4E	Eukaryotic translation initiation factor 4E
eIF4G	Eukaryotic translation initiation factor 4G
ELISA	Enzyme-linked immunosorbent assay
EMS	Ethylmethane sulphonate
EST	Expressed sequence tag
FAO	Food and Agriculture Organization
FAOSTAT	FAO statistical database
FISH	Fluorescence *in situ* hybridization
GC/NPD	Gas chromatography with nitrogen phosphorus detection
gf	*Green-flesh* locus
GM	Genetically modified
GMS	Genic male sterility
GRIN	Germplasm Resources Information Network
HPLC	High performance liquid chromatography
HSCCC	High-speed counter-current chromatography
iaaM	Tryptophan-2-monooxygenase
INDEL	Insertions/deletions
IVC	Institutes of Vegetables Crops, China
ISSR	Inter-simple sequence repeat
LG	Linkage group
LOD score	Logarithm of the odds ratio
LTR	Long terminal repeat
MAB	Marker-assisted backcrossing
MAS	Marker-assisted selection
MIPS	L-Myo-inositol 1-phosphate Synthase
MS	Male sterility
mtlD	Mannitol-1-phosphodehydrogenase
NBPGR	National Bureau of Plant Genetic Resources, India
NGS	Next generation sequencing
NIAS	National Institute of Agrobiological Sciences, Japan
*npt*II	Neomycin phosphtransferase II (gene)
OP	Open-pollinated
ORF	Open reading frame
PaMMV	*Paprika mild mottle virus*
PCR	Polymerase chain reaction
PaO	*Pseudomonas aeruginosa*
PepMoV	*Pepper mottle virus*
PMMoV	*Pepper mild mottle virus*
Pr	Partial restoration

PR	Pathogenesis related
PVMV	*Pepper veinal mottle virus*
PVE	*Potyvirus E*
PVY	*Potato virus Y*
QTL	Quantitative trait loci
RACE	Rapid amplification of cDNA ends
RAPD	Random amplified polymorphic DNA
RDA	Recommended daily amount
RFLP	Restriction fragment length polymorphism
RIL	Recombinant inbred lines
rRAMP	Reverse random amplified microsatellite polymorphism
SAS	Statistical Analysis System
SCAR	Sequence characterized amplified region
SFP	Single feature polymorphism
SGN	SOL Genomics Network
SIM	Simple interval mapping
SNP	Single nucleotide polymorphism
SRAP	Sequence-related amplified polymorphism
SSD	Single-seed decent
SSCP	Single-stranded conformation polymorphism
SSH	Suppressive-subtractive hybridization
SSR	Simple sequence repeat
STS	Sequence tagged site
TEV	*Tobacco etch virus*
TEV-HAT	*Tobacco etch virus-highly aphid transmitted*
TMV	*Tobacco mosaic virus*
ToMV	*Tomato mosaic virus*
TRV	*Tobacco rattle virus*
TSWV	*Tomato spotted wilt virus*
UPOV	International Union for the Protection of New Varieties of Plant
USDA	United States Department of Agriculture
USDA-ARS	USDA-Agricultural Research Service
VIGS	Virus-induced gene silencing
Vpg	Virus genome-linked protein
WGS	Whole genome sequence
YAC	Yeast artificial chromosome
YFP	Yellow fluorescence protein

1

Basic Information on Pepper

Wing Yee Liu,[1] Won-Hee Kang[2] and Byoung-Cheorl Kang[2],*

ABSTRACT

Capsicum species, commonly known as peppers, are among the most important vegetable crops worldwide. The genus includes around 30 species, while *C. annuum*, *C. chinense* and *C. frutescens* are three of the main cultivated species grown for fresh, dried and processing food consumption. As a member of Solanaceae family, *Capsicum* sp. is grouped as the "x = 12" clade of tribe Solanae with closely related important vegetable crops like tomato (*Solanum lycopericum*) and potato (*S. tuberosum*). The origin of *Capsicum* sp. is believed to be South America. It has been domesticated for over thousands of years in Latin America and was introduced in Europe in the 15th century and then distributed worldwide. Wild *Capsicum* plants are perennial shrubs, whereas commercial pepper cultivars are usually grown as annual crops. *Capsicum* is characterized by its biosynthesis of capsaicinoids, which contributes to the pungency. Peppers are also known to have high nutritional values; they contain various antioxidant vitamins, biological pigments, dietary fiber and several minerals that are essential to humans.

Keywords: Pepper consumption, Nutrition, Botanical, Domestication, Capsaicinoid

[1]School of Biological Sciences, The University of Hong Kong, 5N01, Kadoorie Biological Sciences Building, Pokfulam Road, Hong Kong.
[2]Department of Plant Science, Seoul National University, Seoul 151-921, Republic of Korea.
*Corresponding author

1. Economic Importance

1.1 Yield, Production, Import and Export Values, Top Producing Countries

Capsicum species, commonly known as peppers, are among the most important vegetable crops worldwide. According to FAOSTAT 2007 (http://faostat.fao.org/), peppers are grown in all 77 recorded countries with a total production of more than 29 million tons. The world's production of peppers continues to expand with an annual growth rate of 6.7%. India and China are the world's first and second largest producers, respectively, of dry and green peppers. Peppers also play an important role in the world vegetable trade. Worldwide, the import and export values of peppers (dry and green) in 2007 'were about $4.75 and $4.57 billion, respectively (Table 1-1).

Table 1-1 Statistics for peppers by FAO.

		Peppers, dry		Peppers, green
		Ha		Ha
Area Harvested (Worldwide)		1,837,419		1,764,284
		Hg/Ha		Hg/Ha
Yield (Worldwide)		15,121		153,771
		Tonnes		Tonnes
Production Quantity	World	2,778,394	World	27,129,708
	Top countries			
	India	1,200,000	China	14,026,272
	China	250,000	Mexico	1,890,428
	Peru	165,000	Turkey	1,759,224
		USD '000		USD '000
Imports	World	902,069	World	3,844,575
		Tonnes		Tonnes
	World	521,947	World	2,058,703
	Top countries			
	Malaysia	111,737	United States of America	585,025
	United States of America	90,616	Germany	277,774
	Thailand	42,879	United Kingdom	148,082
		USD '000		USD '000
Exports	World	870,970	World	3,699,699
		Tonnes		Tonnes
	World	503,199	World	2,103,699
	Top countries			
	India	220,168	Mexico	530,896
	China	76,665	Netherlands	378,062
	Peru	43,711	Spain	368,534

Source: FAOSTAT 2007

2. Pepper Consumption and Applications

2.1 Food (Fresh, Dried and Processed)

In general, peppers are consumed in two main forms, as food (fresh, dried or processed) or as active ingredients (in industrial and medical products). *Capsicum annuum* is the most prevalent *Capsicum* species cultivated for the fresh vegetable market, although other *Capsicum* species, like *C. chinense* and *C. frutescens,* are also widely grown in various regions (Andrews 1995a). *C. pubescens* and *C. baccatum* have a minor market in Latin American regions. Pepper consumption is closely associated with culture and diet. The cuisines of Thailand, India, Korea, Mexico and several other countries, where peppers are major ingredients, are globally famous for their characteristic spiciness. People in different countries or different regions of a country can have different demands for pepper fruit types, colors, shapes, maturity, taste and pungency. Among all the plant's parts, pepper fruits have the most economic value as a vegetable and an ingredient. The leaves of pepper plants are also consumed as a leafy vegetable in some cultures.

Peppers are also important in the processed food industry. Three species of *Capsicum, C. annuum, C. chinense* and *C. frutescens,* are usually used for processing. Pickles, sauces and powders represent the major processed pepper products of the industry. More than 350 hot sauces from all over the world, including Tabasco (*Capsicum frutescens*), the best-known pepper sauce, were described in 'The Great Hot Sauce Book' (Thompson 1995). Both fresh and dried peppers can be used for making sauces; the famous 'Salsa' sauce, for instance, is made with fresh serrano and jalapeño peppers (Andrews 1995a). To produce pepper powder, also known as paprika, pepper fruits are dried or dehydrated and then processed. Pepper powder is an internationally important spice, particularly in Hungary, India and Korea.

2.2 Active Ingredients (in Industrial and Medical Products)

Capsicum and *Capsicum*-derived ingredients also have extremely diverse applications in traditional medicines, food additives, drugs, health-promoting products, pest control agents in agricultural fields and cosmetic products (as skin-conditioning agents, external analgesics, flavoring agents and fragrance components) (Bosland and Votava 2000). Capsaicinoids have been widely used as topical analgesics as they interact strongly with the mammalian pain receptor, TRPV1 (VR1) (Caterina et al. 2000). Although pain control is the most popular application of these molecules, their various pharmacological applications are reported in hundreds of publications every year.

3. Nutritional Value of Peppers

Peppers are known as a vegetable with high nutritional value. In particular, they contain diverse antioxidant compounds. The nutrient composition varies with pepper species, type, variety and fruit age and status (dried, freeze-dried, raw, pickled, powdered etc.).

3.1 Essential Antioxidant Vitamins

Peppers provide essential antioxidant vitamins to humans. They are particularly rich in lipid-soluble vitamin A and water-soluble vitamin C (Palevitch and Craker 1995). Vitamin A is found in peppers in the form of β-carotene or provitamin A. The accumulation of vitamin A increases as fruits mature and turn red or orange. These high levels of vitamin A are retained in the dried fruits (DeWitt 1999). The quantum of Vitamin C (ascorbic acid) is high in both green fresh fruits and mature fruits (although relatively lower in the mature fruits). Unlike vitamin A, vitamin C tends to break down in dried or dehydrated pepper products. Peppers are also a good source of vitamin B2 (Riboflavin), vitamin B3 (Niacin), vitamin E (alpha-tocopherol), vitamin K and vitamin B6 (Table 1-2).

3.2 Biological Pigments

Carotenoids and anthocyanins are biological pigments in plants that are often found in peppers. They are responsible for the fruit color and also function as antioxidants. Different combinations of anthocyanins and carotenoids affect the color of the immature fruit. Immature green peppers contain carotenoid pigments, namely β-carotene, cryptoxanthin, lutein, lycopene and zeaxanthin. In combination with carotenoid pigments, the anthocyanin dephinidin produces violet to black pigmentation of immature peppers. Although the anthocyanin pigment is degraded concomitantly with fruit ripening, the carotenoids are accumulated (Lightbourn et al. 2008). The red color of some pepper fruits results from the accumulation of different carotenoids in fruit chromoplasts during ripening. The predominant red pigments are capsanthin and capsorubin, whereas the yellow and orange pigments are lutein, β-carotene (provitamin A), zeaxanthin, violaxanthin and antheraxanthin (Buckenhüskes 2003). The carotenoid biosynthetic pathway in pepper fruits and putative correspondence with quantitative and qualitative organ color loci were identified in Solanaceae (Thorup et al. 2000).

Table 1-2 Nutrition values of peppers.

	Peppers, dry per 100g[1]	Peppers, raw per 100g[2]	% Daily Value[3]
Energy (kcal)	280–350	20–40	
Energy (kJ)	1180–1310	84–169	
Protein	11–18	0.8–2.0	
Total lipid, fat (g)	3–16	0.2–0.6	
Ash (g)	6–8	0.4–0.9	
Carbohydrate (g)	51–70	5.4–9.5	
Fiber, total dietary (g)	21–29	0.9–3.7	
Sugars (g)	38–41	2.0–5.3	
Cholesterol	0	0	
Minerals			
Calcium (mg)	45–134	7–18	0.7–1.8
Iron, Fe (mg)	6–11	0.3–1.2	2.7–8
Magnesium, Mg (mg)	88–188	10–25	3.2–8.1
Phosphorus, P (mg)	159–327	20–46	2.9–6.6
Potassium, K (mg)	1870–3170	175–340	8.75–17
Sodium, Na (mg)	43–193	1–13	0.04–0.54
Zinc, Zn (mg)	1–2	0.1–0.3	1.1–2.5
Copper, Cu (mg)	0.2–1.4	0.07–0.13	4.4–8.9
Manganese, Mn (mg)	0.8–1.9	0.10–0.24	4.5–10.8
Selenium, Se (mcg)	2.9–3.7	0.1–2	0.2–3.6
Vitamins			
Vitamin C, total ascorbic acid (mg)	2–31.4	44.3–183.5	73–306
Thiamin (mg)	0.1–1.2	0.03–0.14	2.5–13.1
Riboflavin (mg)	1.2–2.4	0.03–0.09	2.5–8.2
Niacin (mg)	6.4–7.4	0.05–1.2	0.3–8.9
Pantothenic acid (mg)	0.5–2	0.10–0.32	2.0–6.3
	Peppers, dry per 100g[1]	Peppers, raw per 100g[2]	% Daily Value[3]
Vitamin B-6 (mg)	0.8–.5	0.22–0.51	17.2–39.1
Folate, total (mcg)	51–229	10–47	–
Choline, total (mg)	84–89	5.5–11.1	
Vitamin A, RAE (mcg)	1020–3860	10–157	1–15.7
Carotene, beta (mcg)	14840–42890	120–670	2.4–13.4
Carotene, alpha (mcg)	313–6930	15–39	
Cryptoxanthin, beta (mcg)	100–1100	0–50	
Lutein + zeaxanthin (mcg)	5200–5800	51–725	
Vitamin E, alpha-tocopherol (mg)	3–4	0.37–0.69	4.6–8.6
Vitamin K, phylloquinone (mcg)	108–114	4.9–14.3	8.9–26

[1] Peppers, dry include sun-dried hot chili, dried ancho, pasilla, freeze-dried sweet green and red pepper
(Source: USDA Nutrient Data Laboratory, http://www.nal.usda.gov)
[2] Peppers, raw include Hungarian, jalapeno, hot chili, yellow, green and red sweet peppers; 100g ≈ one medium-size sweet green pepper
(Source: USDA Nutrient Data Laboratory, http://www.nal.usda.gov)
[3] Based on Dietary Reference Intakes (DRI) (Source: The Journal of Nutrition, http://jn.nutrition.org/nutinfo/)

3.3 Minerals and other Nutrients

Besides antioxidant vitamins and pigments, peppers also contain dietary fiber and several minerals essential to humans. They are a great source of iron, magnesium, manganese, phosphorus, copper and potassium. On the other hand, peppers contain low levels of cholesterol and sodium. Information regarding nutrient compositions of different peppers is provided by the Chile Pepper Institute (www.chilepepperinstitute.org) and the USDA Nutrient Data Laboratory (www.usda.gov).

4. Capsaicinoids and Pungency

4.1 Pungency and the Scoville Scale

Although pungency is a unique feature of *Capsicum*, not all *Capsicum* species are pungent. *C. annuum*, *C. ciliatum* and *C. chacoense* are examples of the non-pungent forms. The Scoville scale, invented by Wilbur Scoville in 1912, has been used for measuring pungency since many years (DeWitt 1999). In a Scoville Organoleptic Test, pepper extracts are sequentially diluted with sugar water and tested by a panel of tasters. The Scoville unit rating for a pepper is determined by the degree of dilution required to make the pungency barely detectable by the tasters (Kachoosangi et al. 2008). Table 1-3 shows the Scoville scale rating of different types of peppers.

Table 1-3 Scoville scale rating of different types of peppers (Chili Pepper Institute; chileman.org).

Pepper types	Species	Scoville scale
Sweet bell pepper	*C. annuum*	0
Pimento pepper	*C. annuum*	0–500
Paprika chili pepper	*C. annuum*	250–1000
Ancho chili pepper	*C. annuum*	1000–1500
Jalapeno pepper	*C. annuum*	500–5000
Serrano pepper	*C. annuum*	5000–25000
Aji	*C. baccatum*	15000–30000
Tabasco pepper	*C. frutescens*	30000–120000
Cayenne pepper	*C. annuum*	30000–50000
Rocoto	*C. pubescens*	50000–100000
Thai chili pepper	*C. annuum*	50000–150000
Habanero chili pepper	*C. chinense*	100000–350000
Bhut Jolokia	*C. chinense*	1001304

4.2 Biosynthesis of Capsaicinoids

The pungency of a pepper is associated with its biosynthesis of capsaicinoids. Capsaicinoids, which have been found uniquely in peppers, include a family of up to 25 related alkaloid analogs. Capsaicin and dihydrocapsaicin are the two major capsaicinoid compounds, in which concentrations of capsaicin are generally higher than those of dihydrocapsaicin (Antonious and Jarret 2006; Antonious et al. 2009). The total content of capsaicinoids varies widely in different fruits of individual plants at different ages, but the fatty acid pattern of capsaicinoids is uniform for all fruits of one plant (Mueller-Seitz et al. 2008). Capsaicinoids are secreted from the glandular epidermal cells of the placenta and accumulate along the epidermis to form the blister in peppers (Stewart et al. 2007).

The biosynthesis of capsaicinoids is complicated. The *AT3* gene located at the *Pun1* locus encoding a putative acyltransferase determines the presence or absence of capsaicinoids (Stewart et al. 2005). Loss of pungency is associated with a 2.5 kb deletion of *AT3* in *C. annuum* or a four base pair deletion in *C. chinense* (Stewart et al. 2005, 2007). The *capsaicin synthase* (*CS*) gene is responsible for capsaicin biosynthesis by carrying out the condensation reaction between vanillylamine and 8-methyl nonenoic acid. However, the accumulation and composition of capsaicinoids varies dramatically among *Capsicum* species. In addition, the genetic inheritance and regulation of capsaicinoid biosynthesis are still poorly understood (Blum et al. 2003). Figure 1-1 shows the current model of the biosynthetic pathway of capsaicinoids.

4.3 Analytical Technology for Capsaicinoid Compounds

As the Scoville scale is not capable of measuring the components of capsaicinoids, various analytical techniques have recently been developed to measure the pungency level, determine the pepper components and monitor capsaicinoids in pepper fruit extracts (Pruthi 2003; Antonious and Jarret 2006). High performance liquid chromatography (HPLC) is one of the most widely applied techniques (DeWitt 1999). Other methods include gas chromatography with nitrogen phosphorus detection (GC/NPD) (Antonious et al. 2009), high-speed counter-current chromatography (HSCCC) (Peng et al. 2009) and Carbon nanotube-based electrochemical sensors.

Additionally, a spectrophotometric approach has been used as an ISO reference method (Pruthi 2003). Advanced technologies have also been

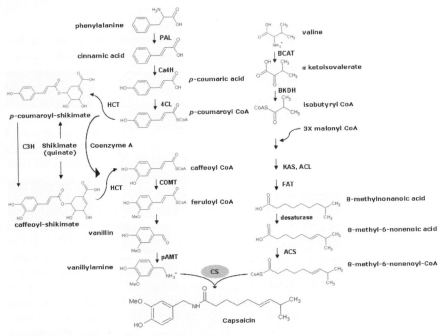

Figure 1-1 Current model of biosynthetic pathway of capsaicinoids.

successfully used to isolate different capsaicinoid compounds, particularly those with the similar structures like capsaicin and dihydrocapsaicin.

5. Evolution and Domestication of *Capsicum*

5.1 Evolution of **Capsicum**

The genus *Capsicum* belongs to a clade comprising tribes Capsiceae (*Capsicum* and *Lycianthes*) and Physaleae in subfamily Solanoideae of family Solanaceae. Subfamilies Solanoideae and Nicotianoideae (including *Nicotiana* and the Australian tribe Anthocercideae) are further grouped as the "$x = 12$" clade (Olmstead et al. 2008). The *Capsicum* species are commonly known as peppers and are closely related to two other important vegetable crops, tomato (*S. lycopersicum*) and potato (*S. tuberosum*), which belong to tribe Solanae (Olmstead et al. 1999; Martins and Barkman 2005) (Fig. 1-2). Pepper and tomato diverged from their last common ancestor approximately 20 million years ago (Wu et al. 2009). The genomes of tomato and pepper are generally well preserved, with 35 conserved syntenic segments in 12 chromosomes 19 inversions, 6 chromosome translocations and a number of single gene transpositions differentiate the two genomes (Wu et al. 2009).

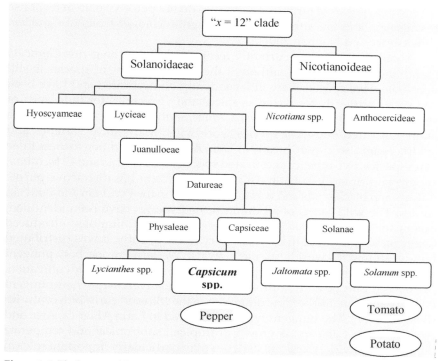

Figure 1-2 Phylogeny of "$x = 12$" clade of the Solanaceae family (Olmstead et al. 2008).

5.2 Domestication of Capsicum

The origin of *Capsicum* species is believed to be in South America, either in central Bolivia along the Río Grande (Andrews 1995b) or in Brazil along the Amazon (DeWitt and Bosland 1993). These are the places where a large number of wild pepper species are concentrated and they are also known as the nuclear areas (DeWitt and Bosland 1993; Andrews 1995b). There are at least two centers of domestication, one in Central America and the other in the Andean region of South America. Capsaicinoids are believed to play an important role in the evolution of *Capsicum* species. Tewksbury et al. (2006) studied the geographic variation of pungency among three species of ancestral *Capsicum* in Bolivia and found that the production of capsaicinoids shifts across elevations. It was suggested that capsaicinoids entail both costs and benefits in response to selection pressure (Tewksbury et al. 2006). It has been hypothesized that this trait evolved to deter mammalian herbivory, wherein the crushing molars and acidic digestion are detrimental to pepper seed survival (seed predator). The favored agents of dispersal for peppers are various species of birds. Avian species do not perceive pain in response to capsaicin and are attracted by the brightly colored fruits of *Capsicum* (Tewksbury and Nabhan 2001).

About 30 species are currently recognized under the genus *Capsicum* (Baral and Bosland 2002), although the actual number of species is still debatable. Five species were independently domesticated and have been cultivated primarily for usage as spices and vegetables for thousands of years (Andrews 1995b). *C. annuum, C. chinense* and *C. frutescens* are the three main cultivated *Capsicum* species worldwide. Many local varieties and landraces are also grown, mostly in Latin America. In addition to these three main species, two other domesticated species, *C. pubescens* and *C. baccatum*, are grown primarily in Latin America. *C. pubescens* has distinctive purple flowers and black seeds and was domesticated in the Peruvian and Bolivian Andes. The wild forms of all domesticated species have been identified, except for that of *C. pubescens* (Pickersgill 1997). Columbus introduced *Capsicum* to Europe in the 15th century. Subsequently, it was distributed rapidly around the world. After its arrival in Western Europe, both pungent and non-pungent forms of the species *C. annuum* came into wide cultivation. In Western Europe and North America, the relatively large, non-pungent bell type (*C. annuum*) became dominant in the 18th and early 19th centuries (Boswell 1937). The pungent types are favored in Latin America, Asia and Africa. Today, *Capsicum* is grown in tropical, subtropical and temperate regions worldwide. *Capsicum* varieties play particularly important roles in the cuisines of several parts of Europe, e.g., Hungary, Latin America, West Africa and many regions of Asia. *Capsicum* is utilized for a diverse range of food products as fresh, dehydrated or processed vegetables and spices. It is also widely used in medicine, in pest and animal control and even in law enforcement, making this a crop of immense cultural and economic importance (Bosland and Votava 2000).

6. Botanical Features

6.1 Botanical Descriptions

Wild *Capsicum* plants are perennial shrubs, whereas commercial pepper cultivars are usually grown as annual crops. As perennial, the herbaceous plant gradually turns woody with age. *Capsicum* species have perfect and complete flowers. As a member of Solanaceae, the flowers of *Capsicum* typically have five sepals, petals, stamens and pistils. The petals (or corolla), stamens and pistils can be white, greenish-white, greenish-yellow or purple. The color combination varies depending on the species and variety. Seed and filament colors, corolla colors and patterns and the number of flowers per node are the keys to identifying the five domesticated *Caspicum* species

(DeWitt and Bosland 1996) (Fig. 1-3, Table 1-4). The cultivated *Capsicum* species are primarily self-pollinated and do not display inbreeding depression. Peppers can also be cross-pollinated by insects, with the outcrossing rate varying by cultivar. Temperatures, particularly the night-time temperatures, are critical for fruit set (Andrews 1995a). The optimal daytime temperatures for fruit set of cultivated peppers are between 20–30°C (chili pepper) and 21–24°C (sweet pepper) but the fruits fail to set or have difficulty in setting when the night temperatures exceed 24°C. Plant growth is also reduced when the air temperature is less than 15°C or more than 32°C (for chili pepper) and less than 18°C or more than 27°C (for sweet pepper) (www.avrdc.org).

6.2 Horticultural Traits of Cultivated Pepper

Pepper fruits are the commodity of the pepper plants; therefore, fruit morphology, flavor and pungency are the most economically important characteristics of *Capsicum*. Major genetic variation is observed with respect to fruit traits such as size, shape, color and flavor, resulting in more than 50 commercially-recognized pod types. Bosland (1992) and Andrews (1995b) described the major fruit types. Types of cultivated *C. annuum* include Ancho/Poblano, Bell, Cayenne, Cherry, Cuban, De Arbol, Exotics, Jalapeño, Mirasol, New Mexican, Ornamentals, Paprika, Pasilla, Pimento, Piquin, Serrano, Squash and Wax (DeWitt and Bosland 1993). The fruit colors before maturity usually range from greenish-white, yellow and green to purple. The mature fruit colors are yellow, orange, red and brown. The majority of green-colored peppers are at the immature fruit stage, but some pepper varieties remain green at maturity. Growth and storage conditions also alter the colors of pepper fruits (Gómez et al. 2003).

The International Union for the Protection of New Varieties of Plants (UPOV, www.upov.int) has published guidelines in four languages for testing the distinctiveness, uniformity and stability (DUS) of *Capsicum annuum* L. (UPOV code: CAPSI_ANN). The guidelines list 54 horticultural characteristics used for DUS testing and the production of consistent

Table 1-4 Flower and seed characteristics of *Capsicum* spp.

	C. pubescens	*C. baccatum*	*C. chinense*	*C. frutescens*	*C. annuum*
Seed Color	Black	Tan	Tan	Tan	Tan
Corolla color	Purple	White	White to Green	Green	White
Corolla pattern	No spot	Spotted	No spot	No spot	No spot
Filament color	Purple	Non-purple	Purple	Non-purple	Non-purple
No. of flowers per node	1	1	>2	1	1

Figure 1-3 Species identification on the basis of morphology. The flower traits are the most distinct between species. All *C. annuum* and *C. chinense* have a white corolla. *C. frutescens* has blue anthers and the corolla color varies from white to greenish-white. The corolla color of *C. baccatum* is white with yellow spots. In *C. pubescens*, both corolla and stamen colors are purple. *C. chacoense* has white and small corollas. Additionally, the seed color of *C. pubescens* accessions is black, while for most *Capsicum* species seed is tan.

Color image of this figure appears in the color plate section at the end of the book.

variety descriptions including plant, leaf, flower and fruit characteristics, lengths of different stages and disease resistance. Table 1-5 summarizes the characteristics established by the UPOV evaluation system.

Table 1-5 Horticultural characteristics for examination.

Items	Characteristics for examination
Plant parts Seedling	Anthocyanin coloration of hypocotyls
Plant	Habit, height, length of stem, shorten internode of upper parts, number of internodes between the first flower and shortened internodes, length of internode (on primary side shoots), Anthocyanin coloration of nodes, intensity of anthocyanin coloration of nodes, hairiness of nodes
Leaf	Length of blade, width of blade, intensity of green color, shape, undulation of margin, blistering, profile in cross section, glossiness
Flower	Peduncle attitude, anthocyanin coloration in anther
Fruit	Color (before maturity), intensity of color (before maturity), anthocyanin coloration, attitude, length, diameter, ratio of length to diameter, shape in longitudinal section, shape in cross section (at level of placenta), sinuation of pericarp at basal part, sinuation of pericarp excluding basal part, texture of surface, color (at maturity), glossiness, stalk cavity, deep of stalk cavity, shape of apex, depth of interloculary grooves, number of locules, thickness of flesh, stalk length, stalk thickness, clayx aspect
Placenta	Capsaicin
Time	Beginning of flowering (first flower on second flowering node), maturity
Resistance	Tobamovirus [Pathotype 0 (*Tobacco mosaic virus* (0)); Pathotype 1-2 (*Tomato mosaic virus* (1–2)); Pathotype 1-2-3 (*Pepper mild mottle virus* (1-2-3))], *Potato virus* Y (PVY) (Pathotype 0, 1, 1–2), *Phytophthora capsici, Cucumber mosaic virus* (CMV), *Tomato spotted wilt virus* (TSWV), *Xanthomonas campestris* pv. *Vesicatoria*

References

Andrews J (1995a) The Domesticated *Capsicum*s. University of Texas Press, Austin, TX, USA.

Andrews J (1995b) Peppers: The Domesticated *Capsicum*. University of Texas Press, Austin, TX, USA.

Antonious GF, Jarret RL (2006) Screening *Capsicum* accessions for capsaicinoids content. J Environ Sci Health. Pt B: Pestic, Food contam Agri Wastes 41: 717–729.

Antonious GF, Berke T, Jarret RL (2009) Pungency in *Capsicum chinense*: variation among countries of origin. J Environ Sci Health, Pt B: Pestic Foodcontam Agri Wastes 44: 179–184.

Baral JB, Bosland PW (2002) An updated synthesis of the *Capsicum* genus. Capsicum Eggplant Newsl 21: 11–21.

Blum E, Mazourek M, O'Connell M, Curry J, Thorup T, Liu K, Jahn M, Paran I (2003) Molecular mapping of capsaicinoid biosynthesis genes and quantitative trait loci analysis for capsaicinoid content in *Capsicum*. Theor Appl Genet 108: 79–86.

**Bosland PW (1993) Breeding for quality in *Capsicum*. Capsicum Eggplant Newsl 12: 25–31.

Bosland PW, Votava EJ (2000) Peppers: Vegetable and spice *Capsicum*s. CABI Publishing, New York, USA.

Boswell VR (1937) Improvement and genetics of tomatoes, peppers, and eggplant. In: USDA [ed] Year Book of Agriculture, Vol 1. Government Printing Office, Washington, USA pp 176–206.

Buckenhüskes HJ (2003) Current requirements on paprika powder for food industry. In: De AK [ed] *Capsicum*: The genus *Capsicum*. Taylor & Francis Inc, New York, USA pp 223–230.

Caterina MJ, Leffler A, Malmberg AB, Martin WJ, Trafton J, Petersen-Zeitz KR, Koltzenburg M, BasbaumAI, Julius D (2000) Impaired nociception and pain sensation in mice lacking the capsaicin receptor. Science 288: 306–313.

Dewitt D (1999) The Chile Pepper Encyclopedia. William Morrow and Co, New York, USA pp 56–57, 118–129, 181–183, 211–214, 219–224.

Dewitt D, Bosland PW (1993) The Pepper Garden. Ten Speed Press. Berkeley, California, USA pp 209.

DeWitt D, Bosland PW (1996) Peppers of the World, An Idenfication Guide. Ten Speed Press, Berkeley, California, USA.

Gómez R, Pardo JE, Alvarruiz A, González M, Varón R (2003) Color differences in peppers and paprikas. In: De AK [ed] *Capsicum*: the genus *Capsicum*. Taylor & Francis Inc, New York, USA pp 236–247.

Kachoosangi RT, Wildgoose GG, Compton RG (2008) Carbon nanotube-based electrochemical sensors for quantifying the 'heat' of chilli peppers: the adsorptive stripping voltammetric determination of capsaicin. Analyst 133: 888–895.

Lightbourn GJ, Griesbach RJ, Novotny JA, Clevidence BA, Raoand DD, Stommel JR (2008) Effects of anthocyanin and carotenoid combinations on foliage and immature fruit color of *Capsicum annuum* L. J Hered 99: 105–111.

Martins TR, Barkman TJ (2005) Reconstruction of Solanaceae phylogeny using the nuclear gene SAMT. Syst Bot 30: 435–447.

Mueller-Seitz E, Hiepler C, Petz M (2008) Chili pepper fruits: content and pattern of capsaicinoids in single fruits of different ages. J Agri Food Chem 56: 12114–12121.

Olmstead RG, Sweere JA, Spangler RE, Bohs L, Palmer JD (1999) Phylogeny and provisional classification of the Solanaceae based on Chloroplast DNA. In: Nee M, Symon DE, Lester RN, Jessop JP [eds] Solanaceae IV: Advances in Biology and Utilization. Royal Botanic Gardens, Kew Publishing, Kew, UK pp 111–137.

Olmstead RG, Bohs L, Migid HA, Santiago-Valentin E, Garciaand VF, Collier SM (2008) A Molecular Phylogeny of the Solanaceae. Taxon 57: 1159–1181.

Palevitch D, Craker LE (1995) Nutritional and medical importance of red pepper (*Capsicum* spp.). J Herbs Spices Med Plants 3: 55–83.

Peng A, Ye H, Li X, Chen L (2009) Preparative separation of capsaicin and dihydrocapsaicin from *Capsicum* frutescens by high-speed counter-current chromatography. J Separ Sci 32: 2967–2973.

Pickersgill B (1997) Genetic resources and breeding of *Capsicum* spp. Euphytica 96: 129–133.

Pruthi JS (2003) Chemistry and quality control of *Capsicum* and *Capsicum* products. In: De AK [ed] *Capsicum*: the genus *Capsicum*. Taylor & Francis Inc, New York, USA pp 25–70.

Stewart C Jr, Mazourek M, Stellari GM, O'Connell M, Jahn M (2007) Genetic control of pungency in *C. chinense* via the *Pun1* locus. J Exp Bot 58: 979–991.

Stewart C Jr, Kang BC, Liu K, Mazourek M, Moore SL, Yoo EY, Kim BD, Paran I, Jahn MM (2005) The *Pun1* gene for pungency in pepper encodes a putative acyltransferase. Plant J 42: 675–688.

Tewksbury JJ, Nabhan GP (2001) Seed dispersal: Directed deterrence by capsaicin in chilies. Nature 412: 403–404.

Tewksbury JJ, Manchego C, Haak DC and Levey DJ (2006) Where did the chili get its spice? Biogeography of capsaicinoid production in ancestral wild chili species. J Chem Ecol 32: 547–564.

Thorup TA, Tanyolac B, Livingstone KD, Popovsky S, Paran I and Jahn M (2000) Candidate gene analysis of organ pigmentation loci in the Solanaceae. Proc Natl Acad Sci USA 97: 11192–11197.

Thompson JT (1995) The Great Hot Sauce Book. Ten Speed Press, New York, USA.

Wu F, Eannetta NT, Xu Y, Durrett R, Mazourek M, Jahn MM and Tanksley SD (2009) A COSII genetic map of the pepper genome provides a detailed picture of synteny with tomato and new insights into recent chromosome evolution in the genus Capsicum. Theor Appl Genet 118: 1279–1293.

2

Classical Genetics and Traditional Breeding in Peppers

Wing Yee Liu,[1] Hee-Bum Yang,[2] Yeong Deuk Jo,[2] Hee-Jin Jeong[2] and Byoung-Cheorl Kang[2,]*

ABSTRACT

Early classical genetic studies in pepper focused on the identification of germplasm and genetic inheritance of important horticultural traits, disease resistance traits and male sterility; desirable traits have been introduced into cultivars through conventional breeding methods and strategies. Major studies and breeding activities have been conducted for resistance to viruses (*Cucumber mosaic virus* and various tobamovirus, potyvirus), bacteria (bacterial leaf spot), fungi (anthracnose) and oomycete (Phytophthora blight). The introgression of disease resistance genes from the wild species or domesticated peppers into the elite breeding materials has contributed significantly to crop improvement. Many horticultural traits such as fruit shape, fruit color, pungency and branching habits have also been studied. Male sterility systems, including cytoplasmic-genic male sterility (CGMS) and genic male sterility (GMS), have been exploited and widely used in commercial F_1 hybrid production; while production of F_1 hybrids by hand emasculation and open-pollinated (OP) varieties are still commonly available. Other tissue culture techniques such as double haploid and embryo culture have been developed for facilitating conventional breeding. The recent development of genetic studies in pepper has been progressed to quantitative trait loci (QTL) analysis and molecular marker development.

Keywords: Germplasm, Conventional breeding, Traits, F_1 Hybrids, Breeding technique

[1]School of Biological Sciences, The University of Hong Kong, 5N01, Kadoorie Biological Sciences Building, Pokfulam Road, Hong Kong.
[2]Department of Plant Science, Seoul National University, Seoul 151-921, Republic of Korea.
*Corresponding author

1. Germplasm

Germplasm collection is extremely important in pepper breeding. The diversity of genetic resources is utilized to add new traits, particularly resistance to new diseases and alternative resistance sources for existing diseases (Fig. 2-1). Besides for crop improvement, germplasm also plays an important role in various types of scientific research, such as elucidating the evolution and classification of *Capsicum* species and in understanding the biology and biochemistry of peppers. The USDA-ARS in Griffin (Georgia, USA) and AVRDC—The World Vegetable Center based in Taiwan has the largest collections of *Capsicum* germplasm and they are active in characterizing, evaluating and curating their collections. AVRDC has a total of 7,726 accessions of *Capsicum* spp. as of September 2006 (www.avrdc.org), and the USDA-ARS has more than 4,700 accessions of 16 *Capsicum* taxa in its active collection program in Griffin (Stoner 2004). These two organizations also maintain databases for their collections and distribute seeds to pepper researchers and breeders all over the world. Other institutions, universities, government or non-government organizations and private seed companies have various scales of *Capsicum* germplasm collections (Berke and Engle 1997). The germplasm of *Capsicum* in at least 15 countries has been reported in the Capsicum and Eggplant Newsletter (CENL) for the past 20 years (1983–2002).

Figure 2-1 Genetic diversity showing various fruit shapes and colors in pepper germplasm. *Color image of this figure appears in the color plate section at the end of the book.*

2. Conventional Breeding and Major Breeding Objectives

There are many types of peppers that are being utilized for different purposes with different quality and trait requirements for successful production. Conventional breeding methods and strategies have been used for crop improvement in pepper, including pedigree breeding, single plant selection by the pedigree method, population improvement by recurrent selection, backcrossing, single-seed descent (SSD) and recombinant inbred lines (RIL) produced by SSD. Disease is often the major constraint of pepper production worldwide. Disease resistance, therefore, is one of the foremost objectives in pepper breeding and genetic studies (Paran et al. 2004). Pohronezny (2003) provided overviews of most of the known pepper diseases. Resistance sources in wild species or domesticated peppers have been reported for *Tomato mosaic virus* (TMV) (Boukema 1980), *Cucumber mosaic virus* (CMV) (Pochard 1982; Shifriss and Cohen 1990), *Potato virus Y* (PVY) (Pochard et al. 1983), *Pepper veinal mosaic virus* (PVMV) (Hobbs et al. 1998), *Tomato spotted wilt virus* (TSWV) (Rosello et al. 1996), bacterial spot (*Xanthomonas campestris* pv. *vesicatoria*) (Hibberd et al. 1983), bacterial wilt (*Ralstonia solanacearum*) (Perera et al. 1992), *Fusarium* wilt (Jones and Black 1992), Anthracnose (*Colletotrichum* spp.) (Voorrips et al. 2004), *Phytophthora* root rot/stem rot/foliar blight caused by *Phytophthora capsici* (Pochard and Daubeze 1982), nematodes, etc.

Disease-resistance breeding and genetics have been studied extensively in the past several years. The introgression of disease resistance genes from the wild germplasm into the elite backgrounds has contributed significantly to crop improvement, particularly in terms of yield and quality enhancement and stability in pepper production. Successful examples are as follows: resistance to TMV from *C. chinense* (L^3) and *C. chacoense* (L^4) (Boukema 1980; Berzal-Herranz et al. 1995; de la Cruz et al. 1997), resistance to TSWV from *C. chinense* (Boiteux et al. 1993) and resistance to bacterial leaf spot disease (*Bs2*) from *C. chacoense* (Cook and Guevara 1984; Kim and Hartmann 1985; Hibberd et al. 1987). These have all been introduced into commercial *Capsicum annuum* cultivars worldwide. Korean breeders developed a CMV-resistant cultivar using the single dominant gene *Cmr1* (Kang et al. 2010) and a *Phytophthora*-resistant cultivar using a conventional breeding method (data not shown). Additional genes for resistance to potyviruses, nematodes and powdery mildew have been identified in several *Capsicum* species and are in use. However, the introgression of disease resistance genes into elite cultivars is particularly difficult when the resistance traits are inherited via a complex quantitative mechanism and linked to undesirable horticultural and economical traits such as low yield and small fruit size. Furthermore, because of recombination and the wide diversity of pathogens, breeding for sustainable resistance remains a difficult task. Therefore, breeding for disease resistance will still be a prioritized objective for pepper breeding in the future. The identification of new resistance sources, breeding for multiple disease resistance and pyramiding different sources of resistance will be important in breeding for sustainable agriculture.

In addition to disease resistance, horticultural and biochemical traits are also important breeding objectives in pepper. The level of pungency is an important and unique aspect of pepper breeding. Understanding people's diet and preferences in terms of the pungency level is particularly significant in pepper breeding. The pungency level is especially important in breeding for the processed food industry. Nowadays, a number of analytical techniques, such as high pressure liquid chromatography (HPLC), have been used to measure the pungency level, determine the components in pepper and monitor capsaicinoids in pepper fruit extracts to assist in pepper breeding and quality control (particularly in industry) (Pruthi 2003; Antonious and Jarret 2006). These analytical techniques are also applied to support breeding for both pungency and color (Pruthi 2003) including fruit color, stability and uniformity of color and color intensity (Bosland 1993). For fresh vegetable peppers, the major traits under selection are related to fruit quality and taste: fruit color, color intensity, size, shape, pericarp thickness, taste and degree of pungency. Furthermore, because of the important role of long-distance vegetable transport, shelf-life has also become an essential breeding objective for fresh vegetable peppers.

3. Traits, Classical Genetics and Molecular Markers

Early classical genetic studies in pepper focused on the identification of germplasm and genetic inheritance of important horticultural traits, disease resistance traits and male sterility. Many useful traits of peppers are inherited via single dominant or recessive genes and quantitative trait loci (QTL). Numerous resistance genes have been studied and reported.

Resistance to tobamovirus is inherited monogenically and is controlled by multiple alleles at the L locus. Boukema (1984) suggested a $L^4 > L^3 > L^2 > L^1 > L^+$ model for the resistance of L alleles, representing the resistance spectrum of the alleles against tobamovirus pathotypes. L alleles affected by temperature changes (Palloix 1992) and a secondary gene stabilizing the resistance conferred by L^1 have been reported (Daubeze et al. 1990). Resistance to *Cucumber mosaic virus* (CMV) varies considerably. Some CMV-resistant genes are inherited dominantly (Kang et al. 2010), some are inherited recessively and others show QTL inheritance (Greenleaf 1986; Herison et al. 2004). The dominant CMV resistance gene is called *Cmr1* whereas the recessive resistance gene is denoted as *cmr2* (Kang et al. 2010). Resistance to potyviruses such as *Chili veinal mottle virus* (CVMV), *Pepper mottle virus* (PepMoV), *Potato virus Y* (PVY), *Potyvirus E* (PVE) and *Tobacco etch virus* (TEV) is controlled by three single dominant genes: *Pvr4*, *Pvr7*, and *Pn1* and five recessive genes: *pvr1*, *pvr3*, *pvr5*, *pvr6* and *pvr8* and QTLs (Wang and Bosland 2006). The *pvr1* locus is responsible for resistance against TEV, PepMoV and PVY (Greenleaf 1956). The *pvr2* locus, which confers resistance to PVY, has several alleles such as $pvr2^1$, $pvr2^2$ and $pvr2^3$ (Kyle and Palloix 1997). An allelism test among *pvr1*, $pvr2^1$ and $pvr2^2$ demonstrated that these three genes are allelic, and they were renamed as *pvr1*, $pvr1^1$ and $pvr1^2$ (Kang et al. 2005). The *pvr3* and *pvr5* genes control resistance to PepMoV and PVY, respectively (Zitter and Cook 1973; Dogimont et al. 1996), whereas *pvr6* and *pvr8* are related to resistance against PVMV and PVY, respectively (Caranta et al. 1996; Arnedo-Andres et al. 2004). *Pvr4* confers the resistance to PepMoV and PVY (Boiteux et al. 1996), and *Pn1* induces a hypersensitive resistance response upon the recognition of PVY (Arnedo-Andres et al. 2004). *Pvr7* controls PVMV resistance and is linked to the *Pvr4* gene (Grube et al. 2000). CVMV is controlled by two recessive resistance genes, *pvr1* and *pvr6* (Hwang et al. 2009). Resistance to TSWV is inherited via a single dominant gene, *Tsw* (Moury et al. 1997). This gene is thermosensitive, and the resistance it confers is not effective at temperatures below 22°C or over 30°C.

Resistance to bacterial leaf spot is conferred by four non-allelic dominant genes: *Bs1*, *Bs2*, *Bs3* and *Bs4* (Kim and Hartman 1985; Hibberd et al. 1987; Sahin and Miller 1997) and two recessive genes, *bs5* and *bs6* (Szarka and Csillery 1995; Csillery et al. 2004). The dominant genes induce the hypersensitive response upon the recognition of *Xanthomonas campestris* in a gene-for-gene manner, whereas the recessive genes do not induce a hypersensitive response. Resistance to *Phytophthra capsici* is inherited via single dominant genes and QTLs. Three dominant genes, *Psr* (Sy 2005), *Pfo* (Walker and Bosland 1999) and *Pfr* (Saini and Sharma 1978) control stem resistance, foliar resistance and fruit rot resistance, respectively. Resistance to anthracnose is conferred by five genes, *Anr1*, *Anr2*, *Anr3*, *Anr4* and *Anr5* or QTLs. Resistance to *Colletotrichum demaitum* is inherited via the *Anr1* gene (Park et al. 1990). *Anr2*, *Anr3*, and *Anr4* are responsible for resistance against *C. gloeosporioides* (Fernandes and Ribeiro 1998). The last gene, *Anr5*, confers resistance against *C. capsici* (Lin et al. 2002). Resistance to powdery mildew is inherited via three genes, *lmr-1*, *lmr-2* and *lmr-3* or QTLs (Shifriss and Pilovski 1992; Lefebvre et al. 2003).

Many horticultural traits such as fruit shape, immature fruit color, mature fruit color, fruit color transition, pungency and branching habits have also been studied. In terms of fruit shape, the pointed fruit apex phenotype is dominantly controlled by the *P* gene and the *Ap* gene (Deshpande 1933; Ishikawa et al. 1998), whereas the non-bulging fruit base shape and enclosed calyx around the fruit base phenotypes are recessively controlled by single genes *fb* and *ce*, respectively (Daskalov and Poulos

1994). Round fruit shape is mainly controlled by dominant gene *O* (Peterson 1959). Two recessive genes, *up-1* and *up-2*, are responsible for the erect fruit phenotype (Gopalakrishnan et al. 1989). The dominant gene *Ped* is related to the acute fruit pedicle. Parthenocarpy is controlled by the recessive *pf* gene (Pathak et al. 1983). Immature fruit colors are inherited via three alleles: sulfur white, yellowish green and cedar green colors. These are conferred by *sw1*, *sw2* and *sw3*, respectively (Odland and Porter 1938). The *sw3* allele is dominant to the other two alleles and *sw2* is dominant to *sw1* (*sw3* > *sw2* > *sw1*). Mature fruit colors are inherited via a series of genes. Two genes, *c-1* and *c-2*, are responsible for the carotenoid synthesis that produces red pigment. Yellow pigment synthesis in fruits is controlled by the *y* gene and is reduced by *c-1* or *c-2* function. The *y+* allele, another allele of the y locus, is responsible for producing red mature fruits. Green color of the mature fruits is a result of chlorophyll retention and is conferred by the *cl* gene. The combination of *y+* and *cl* results in mature fruit being brown. At the intermediate fruit maturation stage, the *im* gene is responsible for purple color. Pungency is mainly determined by the *Pun* gene (previously designated as *C*). *Pun* is responsible for the capsaicinoid synthesis that produces pungency. Branching habits are inherited by several genes. The *bl* gene conditions the branchless habit (Bergh and Lippert 1964). Determinate growth is governed by the *dt* gene and the number of axillary shoots is determined by the *ct* gene. Finally, indeterminate growth is conferred by two dominant alleles, *Dt* and *Ct*, whereas fasciculate branching is controlled by *fa* (Lippert et al. 1965).

Male sterility includes genic male sterility (GMS) and cytoplasmic male sterility (CMS). There are more than 20 genes responsible for GMS. Eight genes were found in mutants by irradiation or ethyl methanesulfonate (EMS) treatment (Daskalov 1973; Daskalov and Poulos 1994). GMS is inherited via recessive genes such as *ms-1*, *ms-2*, *ms-12*, *ms-13* and *ms-14* in spontaneous mutants (Shifriss and Frankel 1969; Shifriss and Rylski 1972; Shifriss 1973; Meshram and Narkhede 1982). By contrast, dominant inheritance of GMS is governed by the *Dms* gene (Daskalov and Poulos 1994). CMS is controlled by the S (sterile) cytoplasm type (Peterson 1958). The nuclear gene, *rf*, can restore the sterility induced by S cytoplasm (Shifriss 1997).

Classical genetics has enabled us to understand the inheritance of useful traits such as disease resistance and horticultural traits. These studies have contributed greatly to pepper crop improvement, particularly through the introgression of disease resistance genes into elite cultivars. In terms of horticultural traits, those which control fruit color (red, yellow and brown) and the presence of pungency are inherited qualitatively, whereas fruit size and shape are a result of quantitative inheritance. Classical genetics makes it possible to study allelic relationships, the determination of dominant/

recessive patterns and epistasis. However, QTL analysis has been limited for a number of these traits.

The availability of molecular resources plays an important role in genetic and genomic studies. Different types of molecular markers have been developed for peppers: isozyme markers (Tanksley 1984), AFLP (amplified fragment length polymorphism), RAPD (random amplified polymorphic DNA), RFLP (restriction fragment length polymorphism), simple sequence repeat (SSR) (Paran et al. 2004; Kang et al. 2005), single nucleotide polymorphism (SNP) (Jung et al. 2010; Yang et al. 2011) and conserved ortholog set (COS) II markers (Wu et al. 2006, 2009). COSII markers, developed from the conserved ortholog regions of tomato and Arabidopsis, belong to the latest generation of molecular markers in pepper. These molecular marker techniques have facilitated great progress in understanding the inheritance of useful traits. Molecular markers have been used to construct pepper genetic maps and perform genetic mapping of horticultural and disease resistance genes and QTLs (Table 2-1).

The development of molecular markers and other genomic techniques also led to the cloning and characterization of genes conferring resistance to potyvirus (*pvr1* and *pvr6*) (Ruffel et al. 2002, 2006; Kang et al. 2005; Hwang et al. 2009), tobomovirus (*L³*) (Tomita et al. 2011) and bacterial spot (*Bs2* and *Bs3*) (Tai et al. 1999a; Tai and Staskawicz 2000; Römer et al 2009). *Fasciculate* (*fa*) (Elitzur et al. 2009) and *Capsaicin synthase* (*csy1*) (Prasad et al. 2006) genes have also been cloned and characterized. Candidate genes have been identified for the presence/absence of pungency (*Pun 1/CS*) (Lee et al. 2005; Stewart et al. 2005), soft flesh and deciduous (*S*) (Rao and Paran 2003) and fruit color (*y, C2, cl*) (Lefebvre et al. 1998; Huh et al. 2001; Efrati et al. 2005).

Even though many genes responsible for useful traits have not been identified, molecular markers to be used for breeding are limited. Molecular markers linked to traits such as resistance to CMV (*Cmr1*) (Kang et al. 2010), TSWV (*Tsw*) (Moury et al. 1997), *Phytophthora* (*Phyto5.2*) (Quirin et al. 2005), and root knot nematode (*Me-1, Me-3, ME-4, Me-7, Mech-1* and *Mech-2*) (Djian-Caporalino et al. 2001; Djian-Caporalino et al. 2007) and the restorer gene (*rf*) (Jo et al. 2010), etc., have been developed and used for marker-assisted selection (MAS) for disease resistance.

4. Other Techniques Used in Pepper Breeding

4.1 Hybrid Seed Production Using Male Sterility

F_1 hybrid (or single-cross hybrid) pepper cultivars are popular in the commercial market, although open-pollinated (OP) varieties are still commonly available. On the basis of heterosis or hybrid vigour, F_1 hybrid peppers generally show significant improvements in yield, plant vigour,

Table 2-1 Summary of classical genetics and genomics studies.

Trait	Locus name	Type of genetic inheritance	Latest status	Representative reference
Horticultural and biochemical traits				
Putative *Acyltransferase* gene AT3 (Presence/Absence of Pungency)	*Pun1*	Gene, Dominant	Mapped, candidate gene	(Lee et al. 2005; Stewart et al. 2005; Stewart et al. 2007)
Fasciculate	*fa*	Gene, Recessive	Mapped, cloned/characterized	(Elitzur et al. 2009)
Fertility restoration	*Rf*	Gene, Dominant	Mapped	(Zhang et al. 2000a; Min et al. 2008; Jo et al. 2009)
	-	QTL	Mapped	(Wang et al. 2004)
Polygalacturonase (Soft flesh and deciduous)	*S*	Gene, Dominant	Mapped, candidate gene	(Rao and Paran 2003)
Capsanthin-capsorubin synthase (CCS) (Yellow fruit color)	*y*	Gene, Recessive	Mapped, candidate gene	(Lefebvre et al. 1998)
Phytoene synthase (Red fruit color)	*C2*	Gene, Dominant	Mapped, candidate gene	(Thorup et al. 2000; Huh et al. 2001)
Chlorophyll retainer (Brown fruit color)	*cl*	Gene, Recessive	Mapped, candidate gene	(Efrati et al. 2005)
Anthocyanin accumulation	*A*	Gene, Dominant	Mapped	(Borovsky et al.2004)
Fruit size	-	QTL	Mapped	(Ben Chaim et al. 2001b; Rao et al. 2003)
Fruit shape	-	QTL	Mapped	(Ben Chaim et al. 2001b; Rao et al. 2003)
Trait	Locus name	Type of genetic inheritance	Latest status	Representative reference
Capsiate	*pAMT*	Gene, Recessive	Candidate gene	(Lang et al. 2009; Tanaka et al. 2010)

Table 2-1 contd....

Table 2-1 contd....

Trait	Locus name	Type of genetic inheritance	Latest status	Representative reference
Male sterility (Genic)	*Camf1*	Gene, Dominant	Candidate gene	Chen et al. 2012
Parthenocarpy	*CaARF8*	Gene, Recessive	Candidate gene	Tiwari et al. 2011
Trichome density	*Pte1*	Gene, Dominant	Map based cloning	Kim et al. 2010
Pungency	*Cap7.1, Cap7.2*	QTL	Mapped	Ben Chaim et al. 2006
Fruit elongation	*fs10.1*	Gene, Dominant	Mapped, Candidated gene	Borovsky et al. 2011
Disease resistance				
Resistance to *Cucumber mosaic virus* (CMV)	*Cmr1*	Dominant	Mapped	Kang et al. 2010
Resistance to *Cucumber mosaic virus* (CMV)	-	QTL	Mapped	(Caranta et al. 1997;Ben Chaim et al. 2001a; Caranta et al. 2002)
Resistance to potyvirus- *tobacco etch virus* (TEV), *potato virus Y* (PVY), *potyvirus E* (PVE)	*pvr1*	Gene, Recessive	Mapped, cloned / characterized	(Murphy et al. 1998; Kang et al. 2005)
Resistance to potyvirus—pepper veinal mottle virus (PVMV)	*Pvr4*	Dominant	Mapped	(Grube et al. 2010)
Resistance to potyvirus-*Chili veinal mosaic virus* (ChiVMV)	*pvr2 (pvr1), pvr6*	Gene, simultaneous double recessive	Mapped; cloned / characterized	(Ruffel et al. 2006; Hwang et al. 2009)
Trait	Locus name	Type of genetic inheritance	Latest status	Representative reference
Resistance to tobamoviruses	L_3	Gene, Dominant	Mapped; cloned / characterized	(Berzal-Herranz et al. 1995; Tomita et al. 2008)

Resistance to tobamoviruses	L_4	Gene, Dominant	Mapped	(Yang et al. 2009)
Resistance to *Tomato spotted wilt virus* (TSWV)	*Tsw*	Gene, Dominant	Mapped	(Jahn et al. 2000; Moury et al. 2000)
Resistance to *Xanthomonas campestris* (Bacterial spot)	*Bs2*	Gene, Dominant	Mapped, cloned/characterized	(Tai et al. 1999b; Mazourek et al. 2009)
Resistance to *Xanthomonas campestris* (Bacterial spot)	*Bs3*	Gene, Dominant	Mapped, cloned/characterized	(Pierre et al. 2000; Jordan et al. 2006)
Resistance to Anthracnose (*Colletotrichum* spp.)	-	QTL	Mapped	(Voorrips et al. 2004)
Resistance to *Phytophthora capsici*	-	QTL	Mapped	(Thabuis et al. 2003; Ogundiwin et al. 2005; Kim et al. 2008)
Resistance to powdery mildew (*Leveillula taurica*)	-	QTL	Mapped	Lefebvre et al. 2003
Resistance to Root-knot nematodes (*Meloidogyne* spp.)	*Me*	Multiple genes, Independent dominant	Mapped	(Djian-Caporalino et al. 2001; Djian-Caporalino et al. 2007)
Resistance to Root-knot nematodes (*Meloidogyne chitwoodi*)	*Mech*	Multiple genes, Independent dominant	Mapped	Djian-Caporalino et al. 2004

number of fruit set, disease resistance, etc. The hybrid cultivars are produced by crossing two homozygous inbred lines, with hand emasculation required to remove the anthers before anthesis of the flowers in female lines; flowers are collected from the male lines and artificial pollination is carried out to produce specific F_1 hybrids. F_1 hybrid pepper cultivars have been produced by seed companies worldwide.

Male sterility, including cytoplasmic-genic male sterility (CGMS) and GMS, has been also used in commercial F_1 hybrid production for a portion of chili pepper cultivars to reduce the cost of production by eliminating the need for emasculation in hybridization and to increase the purity of hybrid seeds (Table 2-2). Of the two types of male sterility, CGMS has been preferred over GMS for hybrid seed production because the segregation of male sterility and male fertility does not occur during the maintenance of male sterile line. In pepper, CGMS was first reported by Peterson (1958) in USDA accession PI164835. No CMS cytoplasm source that is different from Peterson's has been discovered (Shifriss 1997). The anthers of CMS pepper lines produce very little or no viable pollen and are significantly smaller than those of normal lines (Fig. 2-2). Microscopic observation of pollen development in CMS pepper showed the degeneration of tapetum cells in meiosis, which results in the abortion of microspores (Novak et al. 1971; Kim 2011). In the pepper CMS system, male sterility is caused by the presence of a maternally-inherited cytoplasmic factor (S), which has been found in association with an additional abnormal mitochondrial gene *orf507* and a truncated mitochondrial gene *ψatp6-2* (Kim and Kim 2005, 2006; Gulyas et al. 2010) and the simultaneous absence of a nuclear gene for the restoration of fertility.

The induction of male sterility in Arabidopsis by transformation with a portion of *orf507* (*orf456*) and the suppression of expression in the restorer line of pepper imply that *orf507* may be a strong candidate for the gene responsible for CMS (Kim et al. 2007). The dominant *Rf* gene encoded in nuclear DNA is responsible for male fertility restoration. In hybrid seed production, a (S) *rfrf* CMS line (called the A line) serves as the female and is pollinated with a male-fertile restorer line (R line), genotype (N/S) *RfRf*. The A CMS line (A line) is maintained via an isogenic male-fertile maintainer line (B line), genotype (N) *rfrf*, which contains normal cytoplasm (N). The CMS system can give rise to 100 percent male sterility and has been utilized particularly in hybrid seed production. Several molecular markers closely linked to the pepper *Rf* gene have been developed (Zhang et al. 2000; Gulyas et al. 2006; Kim et al. 2006; Min et al. 2008; Jo et al. 2009; Min et al. 2009) to increase the selection efficiency of the breeding lines in the CGMS system. Although these studies showed that the restoration phenotype is determined by a single dominant gene located on the upper region of pepper chromosome six, different inheritance patterns have been reported

Table 2-2 Comparison of CGMS and GMS system for the hybrid seed production in pepper.

Type of male sterility	CGMS	GMS
Localization of male sterility inducing factor	Mitochondrial DNA	Nuclear DNA
Source of male sterility inducing factor	PI164835 (Peterson 1958)	More than dozens of lines in which GMS were induced spontaneously or by mutagenesis using X-ray, gamma ray or EMS
Inheritance of male sterility	Maternally inherited	inherited by a single recessive gene (except for *Dms* which is the single dominant gene)
Inheritance of fertility restoration	Inherited by a single dominant gene (two independent or complementary genes or OTLs were reported in several researches)	Restoration by a dominant wild-type allele of GMS gene
Stability of male sterility and restoration	Stable in a portion of chilli pepper lines including hot dry type pepper in Korea A large portion of hot pepper lines and sweet peppers show unstable male sterility dependant on environmental factors or partial restoration of fertility	Highly stable among most of pepper lines
Range of application	A portion of chilli pepper lines which show stable male sterility and restoration	Mostly applied to pepper lines including sweet peppers and paprikas to which CGMS system is not applicable
Advantage	No segregation during maintenance of male sterile parental line	Highly stable Rapid introgression to elite lines
Disadvantage	Application is limited to a portion of chilli peppers Development of three lines are requisite to hybrid seed production	Segregation during maintenance of male sterile parental line, thus selection process is required and male fertile lines should be removed

in several other studies: the determination of the restoration phenotype by two independent dominant genes (Peterson 1958), the interaction of two complementary genes (Novak et al. 1971) and the involvement of QTL in which one major QTL and four additional minor QTLs were included (Wang et al. 2004).

The pepper CGMS system tends to be more stable in certain types of hot pepper, for instance, the hot dry pepper type in Korea (Shifriss 1997;

 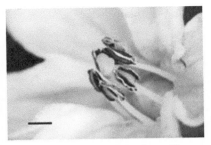

Figure 2-2 Morphology of anthers in fertile (left) and CMS (right) and pepper lines. Figures were quote from Min (2009).

Color image of this figure appears in the color plate section at the end of the book.

Lee 2001), but shows instability in other hot/sweet types, especially at low temperatures (Shifriss and Guri 1979; Shifriss 1997). The plants with unstable sterility are completely sterile under normal growth conditions, but become fertile when the temperature drops below a critical point. In addition to the instability of sterility that is dependent on environmental conditions, the partial restoration (*Pr*) of fertility has been reported (Lee et al. 2008; Min et al. 2009). Pepper plants with *Pr* characteristics contain CMS cytoplasm and produce viable seeds, but they are distinguished from restorer lines in that they bear a mix of normal and aborted pollen sticking to the anther wall, which results in low fruit set and very low seed set per fruit. The *Pr* determinant may be an *Rf* allele or located close to the *Rf* gene. The instability and partial restoration of fertility in sweet peppers and a large number of hot pepper lines hamper the efficient production of hybrid seeds with high purity. In addition, restorer lines are quite common in hot peppers, but rare in sweet peppers (Zhang et al. 2000; Kumar et al. 2001; Yazawa et al. 2002). Pepper types, therefore, restrict the utilization of CMS in hybrid seed production.

GMS is available and has been used for sweet pepper production. In the GMS system, male sterility is conferred by a pair of homozygous recessive genes (*ms/ms*) while homozygous dominant or heterozygous (*Ms/Ms* or *Ms/ms*) plants are male fertile. To maintain and produce a male-sterile line, two isogenic lines with the only difference at the *Ms* locus (*Ms/Ms* and *ms/ms*) are crossed, which gives rise to a progeny mixture of 50 percent male-fertile (*Ms/ms*) and 50 percent male-sterile (*ms/ms*) plants. For hybrid seed production, the male-fertile plants are manually identified and removed. The remaining male-sterile (*ms/ms*) plants are used as females and pollinated with the second parental line (*Ms/Ms*) for hybrid seed production. Although the maintenance of the male-sterile line is less efficient in GMS than in the CMS system, GMS has an advantage in that GMS genes can be introgressed easily by simple backcrosses to new elite lines with diverse genetic backgrounds

(Lee et al. 2008). Dozens of GMS genes have been discovered or induced by mutagenesis using X rays or gamma rays and EMS (Shifriss 1997). Only a few GMS lines, including GMS1, GMS3 and GMSK, have been used for hybrid seed production commercially. Recent studies on the development of markers linked to GMS genes may facilitate the efficient utilization of the GMS system for hybrid seed production in sweet peppers, especially in paprikas (Lee et al.). Both CMS and GMS systems dramatically reduce the cost of production of F_1 hybrid peppers by eliminating the need for emasculation in hybridization.

4.2 Doubled Haploid Technology

Doubled haploid (DH) technology has been used to facilitate conventional breeding programs to produce homozygous inbred lines. Since DH lines are homozygous inbreds ready for crossing to produce hybrids, tremendous time and cost could be saved by this technology compared with the conventional breeding approach in which six to eight generations are required to produce homozygous inbred lines (Khush and Virmani 1996). In addition, DH lines can be used in other aspects of genetic research such as genetic mapping and the evaluation of complex quantitative traits (Thabuis et al. 2003, 2004). In pepper, DH lines can be generated by the induction and regeneration of haploid embryos from anthers or microspores, followed by doubling of the haploid genome to produce diploid plants (Irikova et al 2011).

Embryogenesis in the generation of DH lines in pepper is often hampered by difficulty in regeneration of normal plants. The regeneration of pepper is less efficient than that of model Solanaceae crops such as tobacco, tomato and potato due to several factors including the severely recalcitrant morphogenic nature of pepper, the formation of rosette shoots or ill-defined shoot buds and the dependence of the regeneration capacity on the genotype of the cultivars (Kothari et al. 2010). In addition, induction of microspore cell division and switching from a gametophytic to an embryogenic pathway are key factors for increasing the regeneration efficiency. Therefore, detailed adjustment of embryogenesis conditions has been performed for anther and microspore cultures in numerous studies.

The successful regeneration of pepper anthers was first reported by Wang et al. (1973) and George and Narayanaswamy (1973). A comprehensive protocol for anther culture in pepper was later published in which a number of factors related to the efficiency of pollen production were identified (Dumas de Valuix et al. 1981). Various modifications or improvements of the original protocol have been reported (Qin and Rotino 1993; Mitykó et al. 1995; Mitykó and Juhász 2006). The optimal conditions determined in these studies relate to growth conditions and donor plant age, the genotype of donor plants, microspore developmental stage and composition of the culture media, plant growth regulators, temperature stress treatment and colchicine treatment for genome diploidization (Irikova et al. 2011). *In vitro* haploid plants are generated directly from haploid gametes ($n = x = 12$ in *Capsicum annuum*), and the chromosome number of haploid plants is doubled with colchicine treatment. DH lines with different recombinant traits can be generated by culturing recombined haploid gametes from hybrid plants (Sopory and Munshi 1996).

Other approaches for generating haploid plants including microspore culture and shed-microspore culture have also been studied (Mityko and Fári 1997; Supena et al. 2006). Microspore culture offers an advantage in that the formation of callus and embryos from somatic tissue can be prevented; all of the regenerated plants can be assumed to be haploids or doubled haploids. Kim et al. (2008) reported high-efficiency plant regeneration from microspores by using modified NLNS medium, and Lantos et al. (2009) established an improved protocol for microspore culture in which microspores are co-cultured with ovary tissues of pepper and wheat. Supena et al. (2006) developed a successful shed-microspore culture protocol in which four to seven plants per original flower bud were obtained in certain genotypes of hot peppers. However, these techniques have been less commonly applied than anther culture because they are still time- and labor-intensive and difficult to use (Miyko and Fári 1997; Supena et al. 2006; Irikova et al. 2011).

4.3 Embryo Culture

Embryo culture has been used to overcome the hybridization barrier between distant species, dormancy, slow seed germination and immaturity in seeds (Brown and Thorpe 1995). Hybridization incompatibility usually occurs in crosses between species that belonged to different gene pools of *Capsicum*, but may also occur within the primary gene pool between *C. annuum*, *C. chinense* and *C. frutescens*. As a result, the hybridization may fail or may give rise to weak or sterile F_1 or backcrossed plants. For embryo culture, immature embryos are collected after hybridization and transferred to basal growth medium, and plants are regenerated from the rescued embryos. Although the information on embryo culture in pepper is not complete, successful results were reported in several studies (Kothari et al. 2010). Fari (1986) first used embryo culture to overcome incompatibility in interspecific crosses between *C. annuum* and *C. baccatum*. Bodhipadma and Leung (2003) reported a protocol for the regeneration of plants from immature embryos by which flowers and ripened fruits with viable seeds were obtained. This technique allows researchers and breeders to transfer desirable genes/traits, particularly disease resistance genes, from a related wild species to cultivated peppers. For example, the anthracnose resistance that was found in *C. baccatum* lines could be introgressed into *C. annuum* via the rescue and culture of embryos obtained from interspecific crosses to overcome the post-fertilization genetic barrier between *C. annuum* and *C. baccatum*, which results in embryo abortion (Yoon et al. 2006). Besides embryo culture, other methods including bridge crossing through related species, mixing of pollen, protoplast fusion, etc., have also been used to overcome hybridization barriers.

References

Arnedo-Andres M, Arteaga ML, Ortega RG (2004) New genes related to PVY resistance in *C. annuum* L, 'Serrano Criollo de Morelos-334'. In: Proceedings of the 12th Eucarpia Meeting on Genetics and Breeding of Capsicum and Eggplant, Noordwijkerhout, Netherlands pp 134–138.

Antonious GF, Jarret RL (2006) Screening *Capsicum* accessions for capsaicinoids content. J Environ Sci Health, Pt B: Pestic Food contam Agri Wastes 41: 717–729.

Ben Chaim A, Grube RC, Lapidot M, Jahn M, Paran I (2001a) Identification of quantitative trait loci associated with tolerance to cucumber mosaic virus in *Capsicum annuum*. Theor Appl Genet 102: 213–1220.

Ben Chaim A, Paran I, Grube RC, Jahn M, Wijk van R, Peleman J (2001b) QTL mapping of fruit-related traits in pepper (*Capsicum annuum*). Theor Appl Genet 102: 1016–1028.

Bergh BO, Lippert LF (1964) Six new mutant genes in the pepper, *Capsicum annuum*. L. J Hered 55: 296–300.

Berke T, Engle L (1997) Current status of major *Capsicum* germplasm collections worldwide. Capsicum Eggplant Newsl 16: 80–82.

Berzal-Herranz A, Cruz de la A, Tenllado F, Diaz-Ruiz JR, Lopez L, Sanz AI, Vaquero C, Serra MT, Garcia-Luque I (1995) The *Capsicum L³* gene-mediated resistance against the tobamoviruses is elicited by the coat protein. Virology 209: 498–505.

Bodhipadma K, Leung DWM (2003) *In vitro* fruiting and seed set of *Capsicum annuum* L. cv. Sweet Banana. *In vitro* Cell Dev Biol Plant 39: 536–539.

Boiteux LS, Cupertino FP, Silva C, Dusi AN, Monte-Neshich DC, van der Vlugt RAA, Fonseca MEN (1996) Resistance to potato virus Y (pathotype 1-2) in *Capsicum annuum* and *C. chinense* is controlled by two independent major genes. Euphytica 87: 53–58.

Boiteux LS, Cupertino FP, Reifschneider FJB (1993) *Capsicum* chinense PI159236: a source of resistance to Phytophthora capsici and *tomato spotted wilt virus*. Capsicum Eggplant Newsl 12: 76.

Borovsky Y, Oren-Shamir M, Ovadia R, De Jong W, Paran I (2004) The A locus that controls anthocyanin accumulation in pepper encodes a MYB transcription factor homologous to Anthocyanin2 of Petunia. Theor Appl Genet. 109: 23–29.

Bosland PW (1993) Breeding for quality in *Capsicum*. Capsicum Eggplant Newsl 12: 25–31.

Boukema IW (1980) Allelism of genes controlling resistance to TMV in *Capsicum* L. Euphytica 29: 433–439.

Boukema IW (1984) Resistance to TMV in *Capsicum chacoense* Hunz. is governed by an allele of the *L*-locus. Capsicum eggplant Newsl 3: 47–48.

Brown DCW, Thorpe TA (1995) Crop improvement through tissue culture. World J Microbiol Biotehcnol 11: 409–415.

Caranta C, Palloix A, Gebere-Selassie K, Lefebvre V, Moury B, Daubeze AM (1996) A complementation between two genes originating from susceptible *Capsicum annuum* L. lines confers a new and complete resistance to pepper veinal mottle virus. Phytopathology 86: 739–743.

Caranta C, Palloix A, Lefebvre V, Daubeze AM (1997) QTLs for a component of partial resistance to cucumber mosaic virus in pepper: restriction of virus installation in host-cells. Theor Appl Genet 94: 431–438.

Caranta C, Pflieger S, Lefebvre V, Daubeze AM, Thabuis A, Palloix A (2002) QTLs involved in the restriction of *Cucumber mosaic virus* (CMV) long-distance movement in pepper. Theor Appl Genet 104: 586–591.

Csillery G, Szarka S, Sardi E, Mityko J, Kapitany J, Nagy B, Szarka J (2004) The unity of plant defense: Genetics, breeding and physiology. In: Proceedings of the 12th Eucarpia Meeting on Genetics and Breeding of Capsicum and Eggplant, Noordwijkerhout, Netherlands pp 147–153.

Cook AA, Guevara YG (1984) Hypersensitivity in *Capsicum chacoense* to race 1 of the bacterial spot pathogen of pepper. Plant Dis 68: 329–330.

Daskalov S (1973) Investigation of induced mutants in *Capsicum annuum* L. III. Mutants in the variety Zlaten Medal. Genet Plant Breed 6: 419–429.

Daskalov S, Poulos JM (1994) Updated *Capsicum* gene list. Capsicum Eggplant Newsl 13: 16–26.

Daubeze AM, Palloix A, Pochard E (1990) Resistance of androgenic autodiploid lines of pepper to Phytophthora capsici and tobacco mosaic virus under high temperature. Capsicum Newsl 8/9: 47–48.

de la Cruz A, Lopez L, Tenllado F, Diaz-Ruiz JR, Sanz AI, Vaquero C, Serra MT, Garcia-Luque I (1997) The coat protein is required for the elicitation of the *Capsicum* L_2 gene-mediated resistance against the tobamoviruses. Mol Plant-Microbe Interact 10: 107–113.

Deshpande RB (1933) Studies in Indian chillies. 3. The inheritance of some characters in *Capsicum annuum* L. Indian J Agri Sci 3: 219–300.

Djian-Caporalino C, Pijarowski L, Fazari A, Samson M, Gaveau L, O'Byrne C, Lefebvre V, Caranta C, Palloix A, Abad P (2001) High-resolution genetic mapping of the pepper (*Capsicum annuum* L.) resistance loci *Me3* and *Me4* conferring heat-stable resistance to root-knot nematodes (*Meloiogyne* spp.). Theor Appl Genet 103: 592–600.

Djian-Caporalino C, Fazari A, Arguel MJ, Vernie T, VandeCasteele C, Faure I, Brunoud G, Pijarowski L, Palloix A, Lefebvre V, Abad P (2007) Root-knot nematode (*Meloidogyne* spp.) *Me* resistance genes in pepper (*Capsicum annuum* L.) are clustered on the P9 chromosome. Theor Appl Genet 114: 473–86.

Dogimont C, Palloix A, Daubeze AM, Marchoux G, Gebre Selassie K, Pochard E (1996) Genetics of broad spectrum resistance to potyviruses in haplodiploid progenies of pepper (*Capsicum annuum*). Euphytica 88: 231–239.

Dumas de Valuix R, Chambonnet D, Pochard E (1981) Culture *in vitro* d' anthères du piment (*Capsicum annuum* L.): amélioration des taux d'obtention de plantes chez différents génotypes par des traitements à +35°C. Agronomie 1: 859–864.

Efrati A, Eyal Y, Paran I (2005) Molecular mapping of the chlorophyll retainer (*cl*) mutation in pepper (*Capsicum* spp.) and screening for candidate genes using tomato ESTs homologous to structural genes of the chlorophyll catabolism pathway. Genome 48: 347–351.

Elitzur T, Nahum H, Borovsky Y, Pekker I, Eshed Y, Paran I (2009) Co-ordinated regulation of flowering time, plant architecture and growth by *FASCICULATE*: the pepper orthologue of *SELF PRUNING*. J Exp Bot 60: 869–880.

George L, Narayanaswamy S (1973) Haploid Capsicum through experimental androgenesis. Protoplasma 78: 467–480.

Gopalakrishnan TR, Gopalakrishnan PK, Peter KV (1989) Inheritance of clusterness and fruit orientation in chilli (*Capsicum annuum* L.). Indian J Genet 49: 219–222.

Greenleaf WH (1956) Inheritance of resistance to tobacco etch virus in *Capsicum frutescens* and in *Capsicum annuum*. Phytopathology 46: 371–375.

Greenleaf WH (1986) Pepper breeding. In: Mark JB [ed] Breeding Vegetable Crops. AVI Publishing, Westport, Connecticut, USA pp 67–134.

Grube RC, Blauth JR, Arnedo-Andres M, Caranta C, Jahn MK (2000) Identification and comparative mapping of a dominant potyvirus resistance gene cluster in *Capsicum*. Theor Appl Genet 101: 852–859.

Gulyas GY, Kim, Shin H, Lee JS, Hirata Y (2010) Altered transcript reveals an *Orf507* sterility-related gene in chili pepper (*Capsicum annum* L.). Plant Mol Biol Rep 28: 605–612.

Fernandes R, Ribeiro de LD (1998) Mode of inheritance of resistance in *Capsicum annuum* accessions to *Colletotrichum gloeosporioides*. In: Proceeding of the 10th Eucarpia Meeting on Genetics and Breeding of Capsicum and Eggplant, Avignon, France pp 170.

Fari M (1986) Pepper (*Capsicum annuum* L.). In: Bajaj YPS [ed] Biotechnol Agriculture and Forestry, v.2: Crops, Springer-Verlag Berlin, Heidelberg, New York, USA pp 345–362.

Herison C, Rustikawati S (2004) Genetics of resistance against *cucumber mosaic virus* (CMV) in hot pepper (*Capsicum annuum* L.). Capsicum Eggplant Newsl 23: 113–116.

Hibberd AM, Stall RE, Subtamanya R (1983) Hypersensitive resistance in a *Capsicum annuum* genotype to *Xanthomonas campestris* pv. *vesicatoria* races 1 and 2. Capsicum Eggplant Newsl 2: 121–122.

Hibberd AM, Bassett MJ, Stall RE (1987) Allelism tests of three dominant genes for hypersensitive resistance to bacterial spot of pepper. Phytopathology 77: 1304–1307.

Hobbs HA, Valerde RA, Black LL, Pufresne DJ, Ariyaratne I (1998) Resistance of *Capsicum* spp. genotypes to *pepper mottle potyvirus* isolates from the western hemisphere. Capsicum Eggplant Newsl 17: 46–49.

Huh JH, Kang BC, Nahm SH, Kim S, Ha KS, Lee MH, Kim BD (2001) A candidate gene approach identified phytoene synthase as the locus for mature fruit color in red pepper (*Capsicum spp.*) Theor Appl Genet 102: 524–530.

Hwang J, Li J, Liu WY, An SJ, Cho H, Her NH, Yeam I, Kim D, Kang BC (2009) Double mutations in *eIF4E* and *eIFiso4E* confer recessive resistance to *Chilli veinal mottle virus* in pepper. Mol Cells 27: 329–336.

Irikova T, Grozeva S, Rodeva V (2011) Anther culture in pepper (*Capsicum annuum* L.) *in vitro*. Acta Physiol Plant 33: 1559–1570.

Ishikawa K, Janos T, Nunomura O (1998) Inheritance of the fruit shape at the apex and the peduncle attachment of pepper. Capsicum Eggplant Newsl 17: 30–33.

Jahn M, Paran I, Hoffmann K, Radwanski ER, Livingstone KD, Grube RC, Aftergoot E, Lapidot M, Moyer J (2000) Genetic mapping of the *Tsw* locus for resistance to the Tospovirus *tomato spotted wilt virus* in *Capsicum spp.* and its relationship to the *Sw-5* gene for resistance to the same pathogen in tomato. Mol Plant-Microbe Interact 13: 673–682.

Jo YD, Kim YM, Park MN, Yoo JH, Park M, Kim BD and Kang BC (2010) Development and evaluation of broadly applicable markers for Restorer-of-fertility in pepper. Mol Breed 25: 187–201.

Jones MM, Black LL (1992) Sources of resistance among *Capsicum* spp. to Fusarium wilt of pepper. Capsicum Eggplant Newsl 11: 33–34.

Jordan T, Romer P, Meyer A, Szczesny R, Pierre M, Piffanelli P, Bendahmane A, Bonas U, Lahaye T (2006) Physical delimitation of the pepper *Bs3* resistance gene specifying recognition of the AvrBs3 protein from *Xanthomonas campestris* pv. *vesicatoria*. Theor Appl Genet 113: 895–905.

Jung JK, Park SW, Liu WY, Kang BC (2010) Discovery of single nucleotide polymorphism in *Capsicum* and SNP markers for cultivar identification. Euphytica 175: 91–107.

Kang BC, Yeam I, Frantz JD, Murphy JF, Jahn MM (2005) The *pvr1* locus in *Capsicum* encodes a translation initiation factor eIF4E that interacts with *Tobacco etch virus* VPg. Plant J 42: 392–405.

Kang WH, Huy NH, Yang HB, Kwon JK, Jo SH, Seo JK, Kim KH, Choi D, Kang BC (2010) Molecular mapping and characterization of a single dominant gene controlling CMV resistance in peppers (*Capsicum annuum* L.). Theor Appl Genet 120: 1587–1596.

Khush GS, Virmani SS (1996) Haploids in plant breeding. In: Jain MS, Sopory SK, Veilleus RE [ed] *In vitro* Haploid Production in Higher Plants. Vol 1: Fundamental Aspects and Methods. Kluwer Academic Publishers, Dordrecht, The Netherlands pp 11–34.

Kim BS, Hartmann RW (1985) Inheritance of a gene (*Bs3*) conferring hypersensitive resistance to *Xanthomonas campestris* pv *vesicatoria* in pepper (*Capsicum annuum*). Plant Dis 69: 233–235.

Kim DH, Kim BD (2005) Development of SCAR markers for early identification of cytoplasmic male sterility genotype in chili pepper (*Capsicum annuum* L.). Mol Cells 20: 416–422.

Kim DH, Kim BD (2006) The organization of mitochondrial *atp6* gene region in male fertile and CMS lines of pepper (*Capsicum annuum* L.). Curr Genet 49: 59–67.

Kim DH, Kang JG, Kim BD (2007) Isolation and characterization of the cytoplasmic male sterility-associated *orf456* gene of chili pepper (*Capsicum annuum* L.). Plant Mol Biol 63: 519–532.

Kim HJ, Nahm SH, Lee HR, Yoon GB, Kim KT, Kang BC, Choi D, Kweon OY, Cho MC, Kwon JK, Han JH, Kim JH, Park M, Ahn JH, Choi SH, Her NH, Sung JH, Kim BD (2008) BAC-derived markers converted from RFLP linked to *Phytophthora capsici* resistance in pepper (*Capsicum annuum* L.). Theor Appl Genet 118: 15–27.

Kim M, Jang IC, Kim JA, Park EJ, Yoon M, Lee Y (2008) Embryogenesis and plant regeneration of hot pepper (*Capsicum annuum* L.) through isolated microspore culture. Plant Cell Rep 27: 425–434.

Kim Y (2011) Haplotype analysis of CMS-associated DNA markers and genetic analysis of TCMS in pepper. MSc Thesis, Seoul National University, South Korea.

Kothari SL, Joshi A, Kachhwaha S, Ochoa-Alejo N (2010) Chilli peppers-a review on tissue culture and transgenesis. Biotechnol Adv 28: 35–48.

Kumar S, Rai SK, Banerjee MK, Kalloo G (2001) Cytological mechanisms of male sterility in a nuclear-cytoplasmic line of chilli pepper (*Capsicum annuum* L.). Capsicum Eggplant Newsl 20: 64–67.

Kyle MM, Palloix A (1997) Proposed revision of nomenclature for potyvirus resistance gene in *Capsicum*. Euphytica 97: 183–188.

Lantos C, Juhász AG, Somogy G, Ötvös K, Vági P, Mihály R (2009) Improvement of isolated microspore culture of pepper (*Capsicum annuum* L.) via co-culture with ovary tissues of pepper or wheat. Plant Cell Tiss Org Cult 97: 285–93.

Lee CJ, Yoo E, Shin J, Lee J, Hwang HS, Kim BD (2005) Non-pungent *Capsicum* contains a deletion in the capsaicinoid synthetase gene, which allows early detection of pungency with SCAR markers. Mol Cells 19: 262–267.

Lee DH (2001) Studies on unstable fertility of CGMS (Cytoplasmic-genic male sterility) in *Capsicum annuum* L. PhD thesis. Seoul National University, South Korea.

Lee J, Han JH, An CG, Lee WP, Yoon JB (2010a) A CAPS marker linked to a genic male-sterile gene in the colored sweet pepper, 'Paprika' (*Capsicum annuum* L.). Breed Sci 60: 93–98.

Lee J, Yoon JB, Han JH, Lee WP, Kim SH, Park HG (2010b) Three AFLP markers tightly linked to the genic male sterility ms_3 gene in chili pepper (*Capsicum annuum* L.) and conversion to a CAPS marker. Euphytica 173: 55–61.

Lee J, Yoon JB, Park HG (2008) Linkage analysis between the partial restoration (*pr*) and the restorer-of-fertility (*Rf*) loci in pepper cytoplasmic male sterility. Theor Appl Genet 117: 383–389.

Lefebvre V, Daubeze AM, Rouppe van der Voort J, Peleman J, Bardin M, Palloix A (2003) QTLs for resistance to powdery mildew in pepper under natural and artificial infections. Theor Appl Genet 107: 661–666.

Lefebvre V, Kuntz M, Camara B, Palloix A (1998) The capsanthin-capsorubin synthase gene: a candidate gene for the y locus controlling the red fruit colour in pepper. Plant Mol Biol 36: 785–789.

Lin Q, Kannchana-Udomkarn C, Jaunet T, Mongkolporrn O (2002) Inheritance of resistance to pepper anthracnose caused by *Colletotrichum capsici*. Capsicum Eggplant Newsl 21: 85–88.

Lippert LF, Bergh BO, Smith PG (1965) Gene list for the pepper. J Hered 56: 30–34.

Mazourek M, Cirulli ET, Collier SM, Landry LG, Kang BC, Quirin EA, Bradeen JM, Moffett P, Jahn MM (2009) The fractionated orthology of *Bs2* and *Rx/Gpa2* supports shared synteny of disease resistance in the solanaceae. Genetics 182: 1351–1364.

Meshram LD, Narkhede MN (1982) Natural male sterile mutant in hot chilli (*Capsicum annuum* L.). Euphytica 31: 1003–1005.

Min WK (2009) Molecular genetic analysis and allelic discrimination of the *Restorer-of-fertility* (*Rf*) gene in peppers (*Capsicum annuum* L.). PhD thesis, Seoul National University, South Korea.

Min WK, Kim S, Sung SK, Kim BD, Lee S (2009) Allelic discrimination of the Restorer-of-fertility gene and its inheritance in peppers (*Capsicum annuum* L.). Theor Appl Genet 119: 1289–1299.

Min WK, Lim H, Lee YP, Sung SK, Kim BD, Kim S (2008) Identification of a third haplotype of the sequence linked to the *Restorer-of-fertility* (*Rf*) gene and its implications for male-sterility phenotypes in peppers (*Capsicum annuum* L.). Mol Cells 25: 20–29.

Mitykó J, Juhász AG (2006) Improvement in the haploid technique routinely used for breeding sweet and spice peppers in Hungary. Acta Agron Hung 54: 203–219.

Mitykó J, Anderasfalvya A, Csillery G, Fári M (1995) Anther-culture response in different genotypes and F₁ hybrids of pepper (*Capsicum annuum* L.). Plant Breed 114: 78–80.

Mityko J, Fári M (1997) Problems and results of doubled haploid plant production in pepper (*Capsicum annuum* L.) via anther- and microspore culture. ISHS Acta Hort 447: III International Symposium on *in vitro* Culture and Horticultural Breeding pp 281–288.

Moury B, Palloix A, Selassie KG, Marchou G (1997) Hypersensitive resistance to tomato spotted wilt virus in three *Capsicum chinense* accessions is controlled by a single gene and is overcome by virulent strains. Euphytica 94: 45–52.

Moury B, Pflieger S, Blattes A, Lefebvre V, Palloix A (2000) A CAPS marker to assist selection of *tomato spotted wilt virus* (TSWV) resistance in pepper. Genome 43: 137–142.

Murphy JF, Blauth JR, Livingstone KD, Lackney VK, Jahn MK (1998) Genetic mapping of the *pvr1* Locus in *Capsicum* spp. and evidence that distinct potyvirus resistance loci control responses that differ at the whole plant and cellular levels. Mol Plant-Microbe Interact 11: 943–951.

Novak F, Betlach J, Dubovsky J (1971) Cytoplasmic male sterility in sweet pepper (*Casicum annuum* L.). I. Phenotype and inheritance of male sterile character. Z Pflanzenzucht 65: 129–140.

Odland ML, Porter AM (1938) Inheritance of the immature fruit color of peppers. Proc Amer Soc Hort Sci 36: 647–651.

Ogundiwin EA, Berke TF, Massoudi M, Black LL, Huestis G, Choi D, Lee S, Prince JP (2005) Construction of 2 intraspecific linkage maps and identification of resistance QTLs for *Phytophthora capsici* root-rot and foliar-blight diseases of pepper (*Capsicum annuum* L.). Genome 48: 698–711.

Palloix A (1992) Disease of pepper and perspectives for genetic control. In: Proceedings of the 8th Eucarpia Meeting on Genetics and Breeding on *Capsicum* and Eggplant, Rome, Italy pp 120–126.

Paran I, Rouppe van der Voort J, Lefebvre V, Jahn MM, Landry L, van Schriek M, Tanyolac B, Caranta C, Ben Chaim A, Livingstone K, Palloix A, Peleman J (2004) An integrated genetic linkage map of pepper (*Capsicum* spp.). Mol Breed 13: 251–261.

Park HK, Kim BS, Lee WS (1990) Inheritance of resistance to anthracnose (*Colletotrichum* spp.) in pepper (*Capsicum annuum* L.). II. Genetics analysis of resistance to *Colletotrichum dematium*. J Kor Soc Hort Sci 31: 207–212.

Pathak CS, Singh DP, Deshpande AA (1983) Parthenocarpy in chillies (*Capsicum annuum* L.). Capsicum Newsl 2: 102–103.

Perera KDA, Hartman GC, Poulos JM (1992) Development of screening technique for evaluation of resistance to *Pseudomonas solanacearum*. Capsicum Eggplant Newsl 10: 31–32.

Peterson PA (1958) Cytoplasmically inherited male sterility in *Capsicum*. Amer Nat 92: 111–119.

Peterson PA (1959) Linkage of fruit shape and color genes in *Capsicum*. Genetics 44: 407–419.

Pierre M, Noël L, Lahaye T, Ballvora A, Veuskens J, Ganal M, Bonas U (2000) High-resolution genetic mapping of the pepper resistance locus *Bs3* governing recognition of the *Xanthomonas campestris* pv vesicatora AvrBs3 protein. Theor Appl Genet 101: 255–263.

Pochard E (1982) A major gene with quantitative effect on two different viruses: CMV and TMV. Capsicum Eggplant Newsl 1: 54–56.

Pochard E, Daubeze AM (1982) Comparison of three different sources of resistance to *Phytophthora capsici* in *Capsicum annuum*. Capsicum Eggplant Newsl 1: 59–61.

Pochard E, Gebre Selassie K, Marehoux G (1983) Oligogenic resistance to *potato virus Y* pathotype 1-2 in the line "perennial". Capsicum Eggplant Newsl 2: 137–138.

Pohronezny K (2003) Compendium of Pepper Diseases. American Phytopathological Society, Minnesota, USA.

Poulos JM (1994) Pepper breeding (*Capsicum* spp.): Achievements, challenges and possibilities. Plant Breed Abstr 64: 143–155.

Prasad BC, Kumar V, Gururaj HB, Parimalan R, Giridhar P, Ravishankar GA (2006) Characterization of capsaicin synthase and identification of its gene (csy1) for pungency factor capsaicin in pepper (*Capsicum* sp.). Proc Natl Acad Sci USA 103: 13315–13320.

Pruthi JS (2003) Chemistry and quality control of *Capsicum* and *Capsicum* products. In: De, AK [eds] *Capsicum*: The Genus *Capsicum*. Taylor & Francis Inc, New York, USA pp 25–70.

Qin X, Rotino GL (1993) Anther culture of several sweet and hot pepper genotypes. Capsicum Eggplant Newsl 12: 59–62.

Quirin EA, Ogundiwin EA, Prince JP, Mazourek M, Briggs MO, Chlanda TS, Kim KT, Falise M, Kang BC, Jahn MM (2005) Development of sequence characterized amplified region (SCAR) primers for the detection of Phyto.5.2, a major QTL for resistance to Phytophthora capsici Leon. in pepper. Theor Appl Genet 110: 605–612.

Rao GU, Paran I (2003) Polygalacturonase: a candidate gene for the soft flesh and deciduous fruit mutation in *Capsicum*. Plant Mol Biol 51: 135–141.

Rao GU, Ben Chaim A, Borovsky Y, Paran I (2003) Mapping of yield-related QTLs in pepper in an interspecific cross of *Capsicum annuum* and *C. frutescens*. Theor Appl Genet 106: 1457–1466.

Römer P, Strauss T, Hahn S, Scholze H, Morbitzer R, Grau J, Bonas U, Lahaye T (2009) Recognition of AvrBs3-like proteins is mediated by specific binding to promoters of matching pepper *Bs3* alleles. Plant Physiol 150: 1697–1712.

Rosello S, Diez MJ, Jorda C, Nuez F (1996) Screening of *Capsicum chacoense* accessions for *tomato spotted wilt virus* resistance by mechanical inoculation. Capsicum Eggplant Newsl 16: 58–60.

Ruffel S, Gallois JL, Moury B, Robaglia C, Palloix A, Caranta C (2006) Simultaneous mutations in translation initiation factors eIF4E and eIF(iso)4E are required to prevent pepper veinal mottle virus infection of pepper. J Gen Virol 87: 2089–2098.

Ruffel S, Dussault MH, Palloix A, Moury B, Bendahmane A, Robaglia C, Caranta C (2002) A natural recessive resistance gene against *Potato virus Y* in pepper corresponds to the eukaryotic initiation factor 4E (eIF4E). Plant J 32: 1067–1075.

Ruffel S, Dussault MH, Lesage ML, Moretti A, Palloix A, Daunay MC, Moury B, Bendahmane A, Robaglia C, Caranta C (2004) Involvement of the eukaryotic translation initiation factor eIF4E in Solanacean-Potyvirus interactions. In: Proceedings of the 12th Eucarpia Meeting on Genetics and Breeding of *Capsicum* and Eggplant. Noordwijkerhout, the Netherlands pp 167–170.

Sahin F, Miller SA (1997) A source of resistance in *Capsicum* spp. accessions to pepper race 6 of *Xanthomonas campestris* pv. *vesicatoria* [abstr]. Phytopathology 87: 794–799.

Saini SS, Sharma PP (1978) Inheritance of resistance to fruit rot (*Phytophthora capsici* Leon.) and induction of resistance in bell pepper (*Capsicum annuum* L.). Euphytica 27: 721–723.

Shifriss C (1973) Additional spontaneous male-sterile mutant in *Capsicum annuum* L. Euphytica 22: 527–529.

Shifriss C (1997) Male sterility in pepper (*Capsicum annuum* L.). Euphytica 93: 83–88.

Shifriss C, Rylski I (1972) A male sterile (*ms-2*) gene in 'California Wonder' pepper (*C. annuum*). Hort Science 7: 36.

Shifriss C, Pilovsky M (1992) Studies of the inheritance of mature fruit color in *Capsicum annuum* L. Euphytica 60: 123–126.

Shifriss C, Frankel R (1969) A new male sterility gene in *Capsicum annuum* L. J Amer Soc Hort Sci 94: 385–387.

Shifriss C, Cohen S (1990) Test for resistance to *cucumber mosaic virus* (CMV) in *Capsicum annuum* L. germplasm. Capsicum Eggplant Newsl 8-9: 52–53.

Sopory SK, Munshi M (1996) Anther culture. In: Jain MS and Sopory SK and Veilleus RE [eds] *In vitro* haploid production in higher plants. Volume 1: Fundamental aspects and methods. Kluwer Academic Publishers, Dordrecht, The Netherlands pp 145–176.

Stewart C Jr, Mazourek M, Stellari GM, O'Connell M, Jahn M (2007) Genetic control of pungency in *C. chinense* via the *Pun1* locus. J Exp Bot 58: 979–991.

Stewart C Jr, Kang BC, Liu K, Mazourek M, Moore SL, Yoo EY, Kim BD, Paran I, Jahn MM (2005) The *Pun1* gene for pungency in pepper encodes a putative acyltransferase. Plant J 42: 675–688.

Stoner AK (2004) Preservation and utilization of *Capsicum* germplasm. In: Proceedings of the 17th International Pepper Conference, Florida, USA pp 29.

Supena EDJ, Suharsono S, Jacobsen E, Custers JBM (2006) Successful development of a shed-microspore culture protocol for doubled haploid production in Indonesian hot pepper (*Capsicm annuum* L.). Plant Cell Rep 25: 1–10.

Sy O, Bosland PW, Steiner R (2005) Inheritance of *Phytophthora* stem blight resistance as compared to *Phytophthora* root rot and *Phytophthora* foliar blight resistance in *Capsicum annuum* L. J Amer Soc Hort Sci 130: 75–78.

Szarka J, Csillery G (1995) Defence systems against *Xanthomonas campestris* pv. *vesicatoria* in pepper. In: Proceeding of the 9th Eucarpia Meeting on Genetics and Breeding of *Capsicum* and Eggplant, Budapest, Hungary pp 184–187.

Tai TH, Staskawicz BJ (2000) Construction of a yeast artificial chromosome library of pepper (*Capsicum annuum* L.) and identification of clones from the *Bs2* resistance locus. Theor Appl Genet 100: 112–117.

Tai TH, Dahlbeck D, Stall RE, Peleman J, Staskawicz BJ (1999a) High-resolution genetic physical mapping of the region containing the *Bs2* resistance gene of pepper. Theor Appl Genet 99: 1201–1206.

Tai TH, Dahlbeck D, Clark ET, Gajiwala P, Pasion R, Whalen MC, Stall RE, Staskawicz BJ (1999b) Expression of the *Bs2* pepper gene confers resistance to bacterial spot disease in tomato. Proc Natl Acad Sci USA 96: 14153–14158.

Tanksley SD (1984) Linkage relationships and chromosomal locations of enzyme-coding genes in pepper (*Capsicum annuum* L.). Chromosoma 89: 352–360.

Thabuis A, Palloix A, Pflieger S, Daubeze AM, Caranta C, Lefebvre V (2003) Comparative mapping of *Phytophthora* resistance loci in pepper germplasm: evidence for conserved resistance loci across Solanaceae and for a large genetic diversity. Theor Appl Genet 106: 1473–1485.

Thabuis A, Palloix A, Servin B, Daubeze AM, Signoret P, Hospital F, Lefebvre V (2004) Marker-assisted introgression of 4 *Phytophthora capsici* resistance QTL alleles into a bell pepper lines: validation of additive and epistatic effects. Mol Breed 14: 9–20.

Thorup TA, Tanyolac B, Livingstone KD, Popovsky S, Paran I, Jahn M (2000) Candidate gene analysis of organ pigmentation loci in the Solanaceae. Proc Natl Acad Sci USA 97: 11192–11197.

Tomita R, Sekine KT, Mizumoto H, Sakamoto M, Murai J, Kiba A, Hikichi Y, Suzuki K, Kobayashi K (2011) Genetic basis for the hierarchical interaction between tobamovirus spp. and *L* resistance gene alleles from different pepper species. Mol Plant-Microbe Interact 24: 108–177.

Tomita R, Murai J, Miura Y, Ishihara H, Liu S, Kubotera Y, Honda A, Hatta R, Kuroda T, Hamada H, Sakamoto M, Munemura I, Nunomura O, Ishikawa K, Genda Y, Kawasaki S, Suzuki K, Meksem K, Kobayashi K (2008) Fine mapping and DNA fiber FISH analysis locates the tobamovirus resistance gene *L³* of *Capsicum* chinense in a 400-kb region of R-like genes cluster embedded in highly repetitive sequences. Theor Appl Genet 117: 1107–1118.

Voorrips RE, Finkers R, Sanjaya L, Groenwold R (2004) QTL mapping of anthracnose (*Colletotrichum* spp.) resistance in a cross between *Capsicum annuum* and *C. chinense*. Theor Appl Genet 109: 1275–1282.

Walker SJ, Bosland PW (1999) Inheritance of *Phytophthora* root rot and foliar blight resistance in pepper. J Amer Soc Hort Sci 124: 14–18.

Wang D, Bosland PW (2006) The Genes of *Capsicum*. HortScience 41(5): 1169–1187.

Wang LH, Zhang BX, Lefebvre V, Huang SW, Daubeze AM, Palloix A (2004) QTL analysis of fertility restoration in cytoplasmic male sterile pepper. Theor Appl Genet 109: 1058–1063.

Wang YY, Sun CS, Wang CC, Chien NJ (1973) The induction of pollen plantlets of Triticale and *Capsicum annuum* anther culture. Sci Sin 16: 147–151.

Wu F, Mueller LA, Crouzillat D, Petiard V, Tanksley SD (2006) Combining bioinformatics and phylogenetics to identify large sets of single-copy orthologous genes (COSII) for comparative, evolutionary and systematic studies: a test case in the euasterid plant clade. Genetics 174: 1407–1420.

Wu F, Eannetta NT, Xu Y, Durrett R, Mazourek M, Jahn MM, Tanksley SD (2009) A COSII genetic map of the pepper genome provides a detailed picture of synteny with tomato and new insights into recent chromosome evolution in the genus *Capsicum* Theor Appl Genet 118: 1279–1293.

Yang HB, Liu WY, Kang WH, Jahn M, Kang BC (2009) Development of SNP markers linked to the *L* locus in *Capsicum* spp. by a comparative genetic analysis. Mol Breed 24: 433–446.

Yang HB, Liu WY, Kang WH, Kim JH, Cho HJ, Yoo JH, Kang BC (2011) Development and validation of *L* allele-specific markers in *Capsicum*. Mol Breed Doi: 10.1007/s11032-011-9666-7.

Yazawa S, Yoneda H, Hosokawa M (2002) A new stable and available cytoplasmic male sterile line of *Capsicum*. Capsicum Eggplant Newsl 21: 52–55.

Yoon JB, Yang DC, Do JW, Park HG (2006) Overcoming two post-fertilization genetic barriers in interspecific hybridization between *Capsicum annuum* and *C. baccatum* for introgression of anthracnose resistance. Breed Sci 56: 31–38.

Zhang B, Huang S, Yang G, Guo J (2000) Two RAPD markers linked to a major fertility restorer gene in pepper. Euphytica 113: 155–161.

Zitter TA, Cook AA (1973) Inheritance of tolerance to a pepper virus in Florida. Phytopathology 63: 1211–1212.

3

Molecular Linkage Maps of *Capsicum*

Ilan Paran

ABSTRACT

Genetic mapping in *Capsicum* has been initiated more than 50 years ago using morphological markers for fruit traits. Since then, numerous maps have been constructed in pepper mostly consisting of various types of molecular markers. These maps were instrumental in identification of genomic regions containing traits of interest and development of linked markers for use in marker-assisted selection. Furthermore, use of common markers with other Solanaceae crops, especially tomato, allowed comparative mapping between these species and determining the syntenic relationship among genomes. Early maps were based on mostly restriction fragment length polymorphism (RFLP) markers, while subsequent maps included mostly PCR-based markers such as random amplified polymorphic DNA (RAPD), amplified fragment length polymorphism (AFLP) and simple sequence repeat (SSR) markers. More recent maps include single nucleotide polymorphism (SNP) markers, often derived from next-generation sequencing studies of different parental lines. Until recently, most maps were generated in inter-specific crosses because of lack of sufficient level of DNA polymorphism between closely related parents. However, with advancement of DNA sequencing and generation of large number of SNPs even between closely related parents, the limit in DNA polymorphism is much less critical, allowing the generation of maps consisting of thousands of markers on the basis of intra-specific crosses. Such saturated maps allow efficient identification of genes of interest by "map-based cloning approach".

Keywords: Intra-specific map, Inter-specific map, Comparative map, Integrated map

Institute of Plant Sciences, The Volcani Center, Bet Dagan, Israel.

1. Early Mapping Studies

Genetic maps provide the linear orders of genes along the chromosomes and the distance among them, which is calculated based on the recombination frequencies. Genetic mapping allows determining the linkage relationship between genes and markers and provides the framework for the isolation of genes based on their positions on the map. The early maps were mostly based on the small linkage groups containing a limited number of the genes, which control the simple-inherited traits showing visible and discrete variations, such as color (Paran and Levin 2007). The development of molecular markers allowed the construction of saturated maps representing the entire chromosome set of the organism and in locating the genes controlling the quantitative traits that display continuous variations and often determined by multiple genes.

The most recent gene list of pepper includes 292 genes, mostly the morphological markers identified in the mutants and the disease resistance genes (Wang and Bosland 2006). However, the majority of these genes have not been mapped to date. The early linkage studies between morphological markers in pepper were attempted by Deshpande (1933) who postulated three instances of linkage. However, these associations could not be substantiated further. The first study of the definite linkage was reported by Peterson (1959) who found the linkage among *A*, *O* and *G* loci, controlling purple and yellow immature fruit colors and fruit shapes. The linkage between *A* and *O* was verified many years later and the loci were assigned to chromosome 10 (Ben Chaim et al. 2003a). A set of primary trisomic lines corresponding to different pepper chromosomes allowed the localization of ten morphological mutations to trisomic lines (Pochard 1977; Pochard and Dumas de Vaulx 1982). Most of the trisomic lines could be assigned to specific chromosomes by the subsequent mapping studies that incorporated molecular and morphological markers (Lefebvre et al. 2002). For example, the *MoA* (*Modifier of Anthocyanin*) locus was localized to the trisomic BR which was subsequently identified as chromosome 11 by sharing the linkage with *L* that confers the resistance to *Tobacco mosaic virus* (TMV) (Lefebvre et al. 1995; Ben Chaim et al. 2001).

A limited attempt to map the biochemical markers in pepper was done by Tanksley (1984) who determined the linkage relationship among 14 enzyme-coding genes (isozymes). Furthermore, the breakpoint of a reciprocal translocation in a *C. annuum* x *C. chinense* cross could be determined as linked to two isozyme loci that were subsequently assigned to chromosome 1 (Livingstone et al. 1999). However, as the isozyme loci are limited in number and they exhibit a low level of polymorphism, these markers were abandoned in the subsequent mapping studies that relied mostly on DNA markers. The implementation of DNA-based markers for mapping allowed the construction of maps with numerous markers that are scattered throughout the genome and the assignment of linkage groups to the 12 chromosomes of pepper. To date over 20 maps consisting of the molecular markers in peppers have been published (Table 3-1). These maps allowed the syntenic relationship of the pepper genome with other Solanaceae species to be determined and have been used to identify many important loci controlling the traits for pepper production, such as the disease resistance, the fruit morphology and the chemical composition. Furthermore, these mapping studies enabled the molecular markers to be developed as the tools for the marker-assisted selection and cloning of the genes of interest (Paran et al. 2006; Paran and van der Knaap 2007).

2. Evolution of Marker Types

The first maps in pepper consisting of the molecular markers were based mostly on restriction fragment length polymorphism (RFLP) markers (Tanksley et al. 1988; Lefebvre et al. 1995; Livingstone et al. 1999). RFLP analysis is based on the hybridization of radioactively labeled DNA clones (originated from genomic and cDNA libraries) to the membranes containing the restriction enzymes-digested genomic DNA (Box 1). As most of the RFLP markers were originated and mapped in tomato, the syntenic relationship between the pepper and tomato genomes was assessed. These comparative mapping studies indicated that, in general, the gene repertoire was conserved in pepper and tomato. The considerable rearrangement of the gene orders, however, was found depending on whether the chromosomes were differentiated between the two species by translocations and inversions. Since the RFLP analysis is considered to be labor intensive, in more recent maps, the primarily different types of PCR-based markers were used. These include: RAPD or random amplified polymorphic DNA (RAPD), amplified fragment length polymorphism (AFLP), simple sequence repeat (SSR) and conserved ortholog set (COS) (Lefebvre et al. 2002; Lee et al. 2004; Sugita et al. 2005; Minamiyama et al. 2006; Barchi et al. 2007; Wu et al. 2009). RFLP markers with the known map position are often included in new maps mainly as the anchor loci in orderto align the new linkage groups with older ones.

PCR-based markers can be divided into two groups, single-copy or multi-copy markers. While PCR primers used for the single-copy markers amplify a single locus in the genome, the primers that amplify the multi-copy markers detect the multiple loci. The most common multi-copy markers are RAPD and AFLP marker systems. RAPD primers typically amplify up to 10 DNA fragments that are scored as the dominant markers (i.e., bands are recorded as present or absent and the heterozygotes, therefore, cannot be distinguished from the homozygotes). AFLP primers can amplify up to

---

Box 1 Description of the main marker systems used to construct pepper maps.

RFLP: The RFLP (restriction fragment length polymorphism) marker system is based on the occurrence of the variation in the length of DNA fragments that are produced following the cleavage with the restriction enzymes (Botstein et al. 1980). The alteration of the DNA sequence at the restriction enzyme recognition site by the mutation can result in RFLP that is resolved by the gel electrophoresis and visualized by the Southern hybridization.

SSR: The genomes of higher eukaryotes typically contain microsatellites or simple sequence repeats (SSR), in which the motifs of one to six bases are repeated multiple times. These SSR sequences are highly polymorphic and informative because they tend to vary in repeat numbers between the different genetic backgrounds (Tautz 1989). Typically, the oligonucleotides that flank the SSR sequence are used as the primers in the polymerase chain reaction (PCR) to detect the size variation.

RAPD: The random amplified polymorphic DNA (RAPD) markers are based on the principle that an oligonucleotide, usually of 10 arbitrary nucleotides in length, will base pair with several homologous regions in the genome and prime the amplification of these sites by PCR (Williams et al. 1990). The polymorphism is typically caused by the failure of the primer to anneal to its homologous sequence because of the change in the DNA sequence.

AFLP: The amplified fragment length polymorphism (AFLP) markers are the DNA fragments (80–500 bp) obtained by the restriction enzymes digestion, followed by the ligation of oligonucleotide adapters to the fragments and the selective amplification by PCR (Vos et al. 1995). The PCR-primers consist of a core sequence (part of the adapter), a restriction enzyme-specific sequence and one to three selective nucleotides. The AFLP fragments are separated and visualized by the gel-electrophoresis.

COSII: The Conserved ortholog set (COS) markers are derived from the single-copy orthologous genes that are highly conserved between tomato and Arabidopsis, and are, therefore, useful for comparative mapping (Wu et al. 2006). The COSII set consists of the PCR-based markers in which the primers flank and amplify the polymorphic intron sequences.

SNP: The single nucleotide polymorphism (SNP) markers are the DNA sequence variations that occur when a single nucleotide in the genome sequence is altered. SNP provides the most common sources of the variation that occur among the different genetic accessions. The high throughput technologies that have been developed allow the simultaneous scoring of thousands of markers (Kim and Misra 2007).

---

100 DNA fragments and, therefore, allow scoring of hundreds of loci in a relatively short time. For example, a pepper map consisting of 518 markers, mostly AFLPs and RAPDs, was constructed in just two months (Sugita et al. 2005). However, the dominant nature of these markers and their lack of the locus-specificity which hampers their use in the marker-assisted selection (MAS) and gene cloning studies promoted the use of PCR markers that amplify the single loci, such as SSR (Lee et al. 2004; Ben Chaim et al. 2006; Minamiyama et al. 2006; Yi et al. 2006) and COSII (Wu et al. 2009).

Table 3-1 List and characteristics of pepper maps.

Cross	Pop. type	Pop. size	Marker types	No. of markers	No. of linkage groups	Length (cM)[1]	Reference
Doux des landes (*C. annuum*) x PI 159234 (*C. chinense*)	BC₁	46	RFLP	85	14	nd[1]	Tanksley et al. 1988
NuMex R Naky (*C. annuum*) x PI 159234 (*C. chinense*)	F₂	46	RFLP	192	19	720	Prince et al. 1993
NuMex R Naky (*C. annuum*) x PI 159234 (*C. chinense*)	F2	75	AFLP, RFLP, RAPD	1007	13	1246	Livingstone et al. 1999
TF68 (*C. annuum*) x Habanero (*C. chinense*)	F2	107	RFLP, AFLP	580	16	1320	Kang et al. 2001
Maor (*C. annuum*) x Perennial (*C. annuum*)	F2	180	RFLP, RAPD, AFLP	177	12	1740	Ben Chaim et al. 2001
H3 (*C. annuum*) x Vania (*C. annuum*)	DH	101	AFLP, RFLP, RAPD	543	20	1513	Lefebvre et al. 2002
Perennial (*C. annuum*) x Yolo Wonder (*C. annuum*)	DH	114	AFLP,RAPD, RFLP	630	26	1668	Lefebvre et al. 2002
Yolo Wonder (*C. annuum*) x Criollo de Morelos 334 (*C. annuum*)	F2	151	RFLP, RAPD, AFLP	208	18	685	Lefebvre et al. 2002
Maor (*C. annuum*) x BG 2816 (*C. frutescens*)	BC2	248	RFLP	92	12	1100	Rao et al. 2003
TF68 (*C. annuum*) x Habanero (*C. chinense*)	F2	107	RFLP, SSR	333	15	1762	Lee et al. 2004
Integration of 6 maps	F2, BC1, DH		AFLP, RFLP, RAPD	2262	13	1832	Paran et al. 2004
PSP-11 (*C. annuum*) x PI 201234 (*C. annuum*)	RIL	94	AFLP, RAPD, SSR	144	17	1466	Ogundiwin et al. 2005
Joe E. Parker (*C. annuum*) x Criollo de Morelos 334 (*C. annuum*)	F2	94	AFLP, RAPD, SSR	113	16	1089	Ogundiwin et al. 2005
K9-11 (*C. annuum*) x AC2258 (*C. annuum*)	DH	176	AFLP, RAPD	518	16	1043	Sugita et al. 2005

TF68 (C. *annuum*) x Habanero (C. *chinense*)	F2	107	SSR, RFLP	243	14	2201	Yi et al. 2006
BG2814-6 (C. *frutescens*) x NuMex RNaky (C. *annuum*)	F2	100	AFLP, RFLP, SSR	728	16	1358	Ben Chaim et al. 2006
Manganji (C. *annuum*) x Tongari (C. *annuum*)	DH	117	SSR, AFLP, CAPS, RAPD	374	13	1042	Minamiyama et al. 2006
Yolo Wonder (C. *annuum*) x Criollo de Morelos 334 (C. *annuum*)	RIL	297	RFLP, SSR, AFLP	587	49	1857	Barchi et al. 2007
Integration of 4 maps	F2		RFLP, SSR, AFLP, WRKY rRAMP, STS	854	12	1892	Lee et al. 2009
BG2814-6 (C. *frutescens*) x NuMex RNaky (C. *annuum*)	F2	94	COSII, RFLP, SSR	373	12	1613	Wu et al. 2009

[1] not determined

Several maps have been developed primarily on the basis of SSR markers. Initially a small number of SSRs were developed by searching the limited pepper sequences available in GENBANK and by screening genomic libraries with probes containing SSR motifs (Lee et al. 2004). Subsequently, when pepper expressed sequence tag (EST) database was developed, the *in silico* analysis of more than 10,000 pepper EST sequences revealed 1201 SSRs (Yi et al. 2006). A sub-set of 513 SSRs were amplified, of which 29 percent were polymorphic and 139 were mapped in a *C. annuum* x *C. chinense* cross. The additional set of approximately 600 SSR markers was developed by screening genomic libraries (Minamiyama et al. 2006). 25 percent of these markers were polymorphic and 106 SSRs were mapped in a *C. annuum* intraspecific cross. Additional 489 SSRs, mostly proprietary markers, were mapped in a *C. annuum* x *C. frutescens* cross, which allowed the detection of quantitative trait loci (QTLs) controlling the capsaicinoid content (Ben Chaim et al. 2006). Recently, a new set of SSR markers has been developed by Nagy et al. (2007); however, mapping of the latter markers has not been reported.

The use of the new-generation sequencing technologies that are capable of producing the sequences of hundreds of megabases allow the creation of large databases of the sequence variation that can be exploited to produce the single nucleotide polymorphism (SNP) markers (Lister et al. 2009). The development of the genomic tools, such as whole genome-micro arrays and high-throughput genotyping technologies (Kim and Misra 2007), open the way to develop the maps containing thousands of SNP markers. It is anticipated that the SNP-based marker systems will serve as the main platforms in the future mapping studies because the SNP markers maximize the exploitation of DNA variation as well as the discovery of SNP, and the genotyping technologies are rapidly evolving and becoming more cost-effective. Recently, a whole-genome pepper oligonucleotide array representing more than 30,000 unigenes has been constructed (Ashrafi et al. 2009). Single feature polymorphisms (SFPs), (Zhu and Salmeron 2007) were detected by the differences in the hybridization signals of genomic DNA from 43 *Capsicum* accessions to the oligonucleotides on the array. SFPs commonly result from SNPs or from the small insertions/deletions (INDEL) within the genomic sequence homologous to the oligonucleotide.

3. Types of Mapping Populations

Most of the mapping studies in pepper have been performed using non-fixed populations, such as F_2, BC_1 and BC_2 (Table 3-1). Few fixed populations have been used, including doubled haploids (DH) (Lefebvre et al. 2002; Sugita et al. 2005) and recombinant inbred lines (RIL, Ogundiwin et al. 2005; Barchi et al. 2007). In few cases, the introgression lines for specific chromosome segments were used in the QTL mapping studies (Ben Chaim et al. 2003a,b; Zygier et al. 2005).

The mapping populations were constructed from the intraspecific (*C. annuum*) crosses and from the interspecific crosses (Table 3-1). The intraspecific crosses usually involve distantly related pepper types, e.g., Yolo Wonder, a typical bell type and Criollo de Morelos 334, a serrano-type wild pepper (Barchi et al. 2007). The DNA polymorphism can reach up to 40 percent of the markers in such wide *C. annuum* crosses, which permits mapping without the problems commonly associated with the interspecific populations such as the reduced fertility and the segregation distortion. Furthermore, the high level of phenotypic variation within *C. annuum* permits mapping of many loci controlling the agriculturally important traits. All the permanent DH and RIL populations were constructed in *C. annuum* intraspecific crosses because interspecific F_1 hybrids are reluctant to DH technology. The low pollen fertility that is often observed in the segregating progenies in the interspecific crosses hampers multiple generations of selfing in order to produce RIL.

The interspecific mapping populations involving crossing of *C. annuum* with *C. chinense* or *C. frutescens* have been used to increase the level of DNA polymorphism or when the traits of interest are present in other species. At least six interspecific maps were constructed in *C. annuum* x *C. chinense* crosses (Table 3-1). Till date, three maps on the basis of crosses of *C. annuum* x *C. frutescens* were constructed (Table 3-1). The first is a BC_2 population from a cross of *C. annuum* Maor x *C. frutescens* BG 2816 constructed using the RFLP markers (Rao et al. 2003). The second is an F_2 population from the cross of *C. annuum* NuMex RNaky x *C. frutescens* 14-6, which was constructed using mostly public and proprietary SSR marker sets anchored by RFLP and cloned genes (Ben Chaim et al. 2006). The third map is developed on the basis of the same cross described by Ben Chaim et al. (2006), but was based on COSII markers (Wu et al. 2009).

4. Integrated Maps

To maximize the marker information generated in independent studies, attempts have been made to integrate the data from several maps into a common resource. An integrated genetic map was constructed by merging the segregation data from six mapping populations (Paran et al. 2004). These populations included the maps built from the interspecific F_2 cross reported by Livingstone et al. (1999), an interspecific BC_1 cross of *C. annum* x *C. chinense* (Tanyolac and Paran, unpublished) and four intraspecific *C. annuum* maps (Ben-Chaim et al. 2001; Lefebvre et al. 2002). The integrated map included a total of 2,262 loci, of which 320 markers were common to at least two populations and served as the anchor markers. The map covers

1,832 cM distributed in 13 linkage groups. The map integration improved the average marker density throughout the genome to one marker per 0.8 cM. However, 15 gaps of at least 10 cM between the adjacent markers still remain in the map because of the uneven marker distribution. Despite the use of numerous markers, a few small unlinked linkage groups still remained, indicating that the pepper map is still not complete and/or the parents involved in mapping crosses differ by the undefined chromosomal rearrangements.

An integrated map on the basis of four F_2 mapping populations (SNU3, SNU4, SNU5 and SNU6) was constructed in the laboratory of Prof. Byung-Dong Kim, Seoul National University (Lee et al. 2009). SNU3 and SNU6 are based on an interspecific cross of *C. annuum* (cv. TF68) and *C. chinense* (cv. Habanero), while SNU4 and SNU5 are based on the intraspecific *C. annuum* cross of CM334 × cv. Chilsungcho. Various types of markers have been used: AFLP, RFLP, SSR, reverse random amplified microsatellite polymorphism (rRAM) (PCR-based markers in which one anchored primer is based on an SSR motif and the second primer is an arbitrary one) (Min et al. 2008), sequence tagged site (STS) derived from bacterial artificial chromosome (BAC) end sequences and RFLPs and WRKY (based on the conserved sequence of DNA binding motif of the known pepper and tomato WRKY genes) (Kim et al. 2008). The map integration with the common markers was performed by JoinMap 3.0 program. Because of the differentiation of chromosomes 1 and 8 in the cultivated and wild parents, the maps of these chromosomes were kept separated in the intraspecific and interspecific crosses. A total of 854 markers in 12 chromosomes were included in the integrated map. The intra and interspecific maps covered the distances of 1,892 and 1,858 cM, with an average marker density of 2.5 and 2.2 cM, respectively. The main advantage of this integrated map is that it contains many single copy markers (RFLP, SSR and STS). Furthermore, most of the markers differ from the marker sets included in the integrated map of Paran et al. (2004) and in the COSII map of Wu et al. (2009), increasing significantly the marker reservoir of peppers.

A consensus map of *C. annuum* was created by aligning the maps from three *C. annuum* crosses (Levebvre et al. 2002). While the individual maps ranged from 685 to 1,668 cM and consisted of from 16 to 20 linkage groups, the consensus map resulted in 14 linkage groups that could be assigned to the 12 pepper chromosomes. Furthermore, this map contained 100 known-function genes that can be used as the candidates for controlling various traits.

5. Comparative Maps

Although the early genetic maps of pepper contained the markers that were derived from tomato, their small number did not permit a meaningful comparison with the tomato genome. The first comprehensive comparative map of pepper was developed by Livingstone et al. (1999). This map included more than a thousand loci including many of the pepper origin (RFLP, RAPD and mostly AFLP markers). In addition, hundreds of loci were detected by RFLP with the tomato probes, allowing the considerably improved resolution in the assessment of the syntenic relationships between tomato and pepper. This study revealed that the macrosynteny has generally been retained since the divergence of these genera, and that 98 percent of the tomato genome and 95 percent of the pepper genome can be accounted by about 20 relatively large linkage blocks. Since the publication of this study, additional departures from synteny have been identified, often as a consequence of the small internal rearrangements within an otherwise conserved genome segment. Because of at least one reciprocal translocation between *C. annuum* and *C. chinense*, chromosomes 1 and 8 involved in the translocation were merged into a single linkage group in this population. The putative centromeric regions of the linkage groups were marked by clusters of the markers as a result of the reduced recombination in these regions.

The recent map of pepper on the basis of COSII markers provided new insights into the evolution of the cultivated species and on the syntenic relationship with tomato (Wu et al. 2009). The map was constructed from an interspecific cross between *C. annuum* cv. NuMex RNaky and the wild *C. frutescens* accession BG 2814-6. Two hundred ninety-nine orthologous markers, mostly COSII markers that have been previously mapped in tomato, allowed the coverage of 1613 cM and the establishment of 12 linkage groups corresponding to the 12 pepper chromosomes (Fig. 3-1).

On the basis of COSII map and the previous cytological studies, a model describing a reciprocal translocation between chromosomes 1 and 8 has been suggested that occurred in the ancestral wild *Capsicum* genome (shared by wild *C. annuum*, *C. chinense* and *C. frutescens*), which differentiated the cultivated *C. annuum* from the wild species. According to this model, illegitimate paring between the clusters of ribosomal genes (R45S) in the wild metacentric non-homologous chromosomes 1 and 8 resulted in the reciprocal translocation (Fig. 3-2). The result of this translocation is the creation of acrocentric chromosome 8 and submetacentric chromosome 1 in the cultivated *C. annuum*.

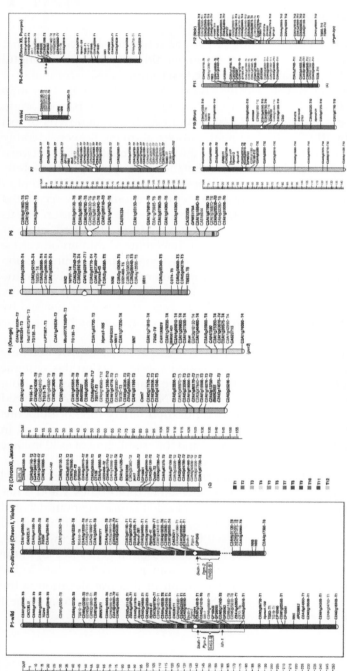

Figure 3-1 A genetic map of pepper based on the COSII markers. The pepper linkage groups are designated P1–P12 based on the synteny with tomato chromosomes and are based on the chromosome names from the trisomic analysis of Pochard (1977). Each tomato chromosome is assigned with one color according to the color code below P2 and the corresponding pepper chromosomes are shown with the same color. ~TN following pepper markers indicates their corresponding tomato chromosomes. The putative centromeric regions are indicated by the white dot. More details on this map can be found in Wu et al. (2009). [Reprinted from Wu et al. (2009) by permission.]

Color image of this figure appears in the color plate section at the end of the book.

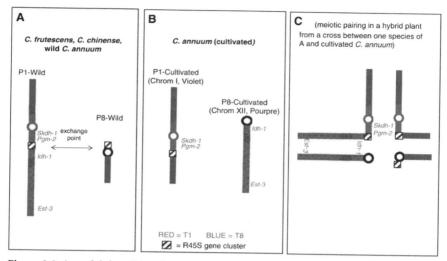

Figure 3-2 A model describing the translocation between the cultivated *C. annuum* and the wild *C. annuum* and the related *Capsicum* species *C. chinense* and *C. frutescens*. [Reprinted from Wu et al. (2009) by permission.]

Color image of this figure appears in the color plate section at the end of the book.

Comparing the pepper and the tomato maps enabled the syntenic relationship between the two genomes to be inferred. 35 syntenic segments ranging from 6 to 117 cM that cover 69 percent of the pepper map were found. A minimum of 19 inversions and five reciprocal translocations differentiates the pepper and tomato genomes (Fig. 3-3). The additional mechanisms, such as the transpositions, must be accounted to explain the positions of the single markers in the non-homologous chromosomes. Chromosome 1 of the wild pepper was the largest chromosome (252 cM), which corresponds to tomato chromosome 1 and most of tomato chromosome 8. The wild pepper chromosome 8 was the smallest chromosome (only 26 cM) that corresponds to a small part of tomato chromosome 8. All other pepper chromosomes except for chromosomes 2, 6 and 7 were also involved in the translocation events that differentiate pepper and tomato genomes.

6. Future Perspectives

The development of the new sequencing and genotyping technologies will accelerate the rate of the discovery of the sequence variation and the map production by high-throughput marker genotyping in the future. The future maps will rely mostly on SNP markers that will allow the construction of the ultra-high density maps containing thousands of markers. The high-density maps will aid the rapid construction of the high-resolution maps

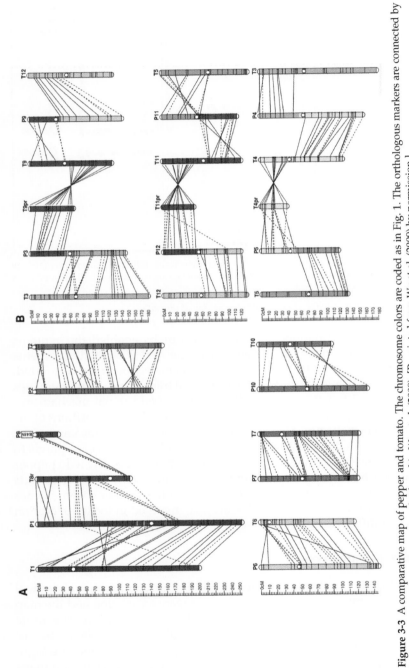

Figure 3-3 A comparative map of pepper and tomato. The chromosome colors are coded as in Fig. 1. The orthologous markers are connected by lines. More details on this map can be found in Wu et al. (2009). [Reprinted from Wu et al. (2009) by permission.]

Color image of this figure appears in the color plate section at the end of the book.

of the targeted regions and positional cloning of the genes of interest. Other applications of the plant genomics utilizing SNP markers such as the whole-genome association studies will become more feasible as the cost of SNP discovery and genotyping will continue to come down. Several mapping studies in pepper have been already targeted the simply and quantitatively-inherited traits related to the disease resistance and the plant development. However, many more studies are needed to ultimately allow the identification of all loci controlling the traits of the agronomic interest and their allelic variation in the wide germplasm and their utilization for the pepper improvement. The acceleration of the gene and marker discovery will have a great impact on the pepper improvement. The new concepts such as "breeding by design" (Peleman and van der Voort 2003) can become reality to push forward the utilization of the marker technologies to increase the efficiency of plant breeding.

References

Ashrafi H, Hill T, Jigiang JY, van Leeuwen V, Michelmore R, Kozik A, van Deynze A (2009) The application of a whole genome pepper array to identify SFPs in a diversity panel. In: Plant and Animal Genome XVII Conference, San Diego, CA, USA, W459.

Barchi L, Bonnet J, Boudet C, Signoret P, Nagy I, Lanteri S, Palloix A, Lefebvre V (2007) A high-resolution, intraspecific linkage map of pepper (*Capsicummannuum* L.) and selection of reduced recombinant inbred line subsets for fast mapping. Genome 50: 51–60.

Ben Chaim A, Paran I, Grube R, Jahn M, van Wijk R, Peleman J (2001) QTL mapping of fruit related traits in pepper (*Capsicum annuum*). Theor Appl Genet 102: 1016–1028.

Ben Chaim A, Borovsky E, De Jong W, Paran I (2003a) Linkage of the *A* locus for the presence of anthocyanin and *fs10.1*, a major fruit-shape QTL in pepper. Theor Appl Genet 106: 889–894.

Ben Chaim A, Borovsky E, Rao GU, Tanyolac B, Paran I (2003b) *fs3.1*: a major fruit shape QTL conserved in *Capsicum*. Genome 46: 1–9.

Ben Chaim A, Borovsky Y, Rao GU, Gur A, Zamir D, Paran I (2006) Comparative QTL mapping of fruit size and shape in tomato and pepper. Israel J Plant Sciences 54: 191–203.

Deshpande RB (1933) Studies in Indian chillies. 3. Inheritance of some characters in *Capsicum*. L. Indian J Agri Sci 3.

Kang B, Nahm SH, Huh JH, Yoo HS, Yu JW, Lee MH, Kim BD (2001) An interspecific (*Capsicum annuum* x *C. chinense*) F$_2$ linkage map in pepper using RFLP and AFLP markers. Theor Appl Genet 102: 531–539.

Kim HJ, Lee HR, Han JH, Yeom SI, Harn CH, Kim BD (2008) Marker production by PCR amplification with primer pairs from conserved sequences of WRKY genes in chili pepper. Mol Cells 25: 196–204.

Kim S, Misra A (2007) SNP genotyping: technologies and biomedical applications. Annu Rev Biomed Eng 9: 289–320.

Lee JM, Nahm SH, Kim YM, Kim BD (2004) Characterization and molecular genetic mapping of microsatellite loci in pepper. Theor Appl Genet 108: 619–627.

Lee HR, Bae IH, Park SW, Kim HJ, Min WK, Han JH, Kim KT, Kim BD (2009) Construction of an integrated pepper map using RFLP, SSR, CAPS, AFLP, WRKY, rRAMP and BAC end sequences. Molecules and Cells 27: 21–37.

Lefebvre V, Pflieger S, Thabuis A, Caranta C, Blattes A, Chauvet JC, Daubeze AM, Palloix A (2002) Towards the saturation of the pepper linkage map by alignment of three intraspecific maps including known-function genes. Genome 45: 839–854.

Lefebvre V, Palloix A, Caranta C, Pochard E (1995) Construction of an intra-specific integrated linkage map of pepper using molecular markers and doubled-haploid progenies. Genome 38: 112–121.

Lister R, Gregory BD, Ecker JR (2009) Next is now: new technologies for sequencing of genomes, transcriptomes and beyond. Current Opinion in Plant Biology 12: 107–118.

Livingstone KD, Lackney VK, Blauth JR, van Wijk R, Jahn MK (1999) Genome mapping in *Capsicum* and the evolution of genome structure in the Solanaceae. Genetics 152: 1183–1202.

Min WK, Han JH, Kang WH, Lee HR, Kim BD (2008) Reverse random amplified microsatellite polymorphism reveals enhanced polymorphisms in the 3′ end of simple sequence repeats on pepper genome. Mol Cells 26: 250–257.

Minamiyama Y, Tsuro M, Hirai (2006) An SSR-based linkage map of *Capsicum annuum*. Mol Breeding 18: 157–169.

Nagy I, Stagel A, Sasvari Z, Roder M, Ganal M (2007) Development, characterization, and transferability to other Solanaceae of microsatellite markers in pepper (*Capsicum annuum* L.). Genome 50: 668–688.

Ogundiwin EA, Berke TF, Massoudi M, Black LL, Huestis G, Choi D, Lee S, Prince JP (2005) Construction of 2 intraspecific linkage maps and identification of resistance QTLs for *Phytophthora capsici* root-rot and foliar-blight diseases of pepper (*Capsicum annuum* L.). Genome 48: 698–711.

Paran I, Rouppe van der Voort J, Lefebvre V, Jahn M, Landry L, van Schriek M, Tanyolac B, Caranta C, Ben Chaim A, Livingstone K, Palloix A, Peleman J (2004) An integrated genetic linkage map of pepper (*Capsicum* spp.). Mol Breed 13: 251–261.

Paran I, Ben Chaim A, Kang BC, Jahn M (2006) *Capsicum* In: Kole C [ed] Genome Mapping & Molecular Breeding in Plants. Vol 5: Vegetables. Springer, Heidelberg, Berlin, New York, Tokyo, pp 209–226.

Paran I, Levin I (2007) Molecular Mapping of Simple Inherited Traits. In: Kole C, Abbott AG [eds] Principles and Practices of Plant Genomics. Vol 1: Genome Mapping. Science Publishers, Inc, Enfield, NH, USA, pp 139–174.

Paran I, van der Knaap E (2007) Genetic and molecular regulation of fruit and plant domestication traits in tomato and pepper. Journal of Exp Bot 58: 3841–3852.

Peleman JD, van der Voort JR (2003) Breeding by design. Trends Plant Sci 8: 330–334.

Peterson PA (1959) Linkage of fruit shape and color genes in *Capsicum*. Genetics 44: 407–419.

Pochard E (1977) Localization of genes in *Capsicum annuum* L. by trisomic analysis. Ann Amelior Plantes 27: 255–266.

Pochard E, Dumas de Vaulx R (1982) Localization of $vy2$ and fa genes by trisomic analysis. Capsicum Newsl 1: 18.

Prince JP, E Pochard, SD Tanksley (1993) Construction of molecular linkage map of pepper and a comparison of synteny with tomato. Genome 36: 404–417.

Rao GU, Ben Chaim A, Borovsky E, Paran I (2003) Mapping of yield related QTLs in pepper in an inter-specific cross of *Capsicum annuum* and *C. frutescens*. Theor Appl Genet 106: 1457–1466.

Sugita T, Kinoshita T, Kawano T, Yuji K, Yamaguchi K, Nagata R, Shimizu A, Chen L, Kawasaki S, Todoroki A (2005) Rapid construction of a linkage map using high-efficiency genome scanning/AFLP and RAPD, based on an intraspecific, doubled-haploid population of *Capsicum annuum*. Breeding Science 55: 287–295.

Tanksley SD (1984) Linkage relationships and chromosomal locations of enzyme-coding genes in pepper (*Capsicum annuum* L.). Chromosoma 8: 352–360.

Tanksley SD, Bernatzky R, Lapitan NL, Prince JP (1988) Conservation of gene repertoire but not gene order in pepper and tomato. Proc Natl Acad Sci USA 85: 6419–6423.

Wang D, Bosland PW (2006) The genes of *Capsicum*. HortScience 41: 1169–1187.

Wu F, Eannetta NT, Xu Y, Durrett R, Mazourek M, Jahn MM, Tanksley SD (2009) A COSII genetic map of the pepper genome provides a detailed picture of synteny with tomato and new insights into recent chromosome evolution in the genus *Capsicum*. Theor Appl Genet 118: 1279–1293.

Yi G, Lee JM, Lee S, Choi D, Kim BD (2006) Exploitation of pepper EST-SSRs and an SSR-based linkage map. Theor Appl Genet 114: 113–130.

Zhu T, Salmeron J (2007) High-definition genome profiling for genetic marker discovery. Trends Plant Sci 12: 196–202.

Zygier S, Ben Chaim A, Efrati A, Kaluzky G, Borovsky Y, Paran I (2005) QTLs mapping for fruit size and shape in chromosomes 2 and 4 in pepper and a comparison of the pepper QTL map with that of tomato. Theor Appl Genet 111: 437–445.

4

Candidate Gene Approaches in *Capsicum*

Michael Mazourek* and Lindsay E. Wyatt

ABSTRACT

The candidate gene approach uses a hypothesis-based method of gene identification. It assumes that a phenotype is caused by sequence polymorphism in a gene rationally associated with the trait. This approach can be contrasted with the map-based cloning approach, which uses molecular markers to identify recombination events between the phenotype and flanking markers to progressively narrow the interval until it contains only the gene of interest. Candidate genes can be compiled based on gene expression data, predicted gene function and/or known position in the genome. Once candidate genes have been identified, they must be tested and validated to be conclusively linked to the phenotype. Several methods are available for validation. In *Capsicum*, direct transgenic complementation is difficult due to the lack of a practical transformation system, but it is possible to transform other solanaceous plants with candidate genes to confirm their function. Virus-induced gene silencing can also be used to validate candidate genes. Finally, co-segregation between the candidate gene and the observed phenotype can support, but not prove, the identity of a candidate gene. The candidate gene approach is particularly suited to use in pepper because of its large genome size and the ability to utilize the wealth of tomato genomic data through comparative approaches. The candidate gene approach has been successfully used in peppers for multiple traits, including the *L* locus for tobamovirus resistance, the *pvr1* and *pvr2* loci for potyvirus resistance, the *Pun1* locus for pungency and the *A* and *cl* loci for fruit color. We believe that the candidate gene approach will continue to be an efficient way to identify genes in the

Department of Plant Breeding and Genetics, Cornell University, Ithaca, NY 14853.
*Corresponding author

future, especially when used to analyze the multiple open reading frame gene candidates generated by next generation sequencing of genomic regions of interest.

Keywords: Candidate gene approach, Ortholog, Capsaicin, Carotenoid, Disease resistance

1. Introduction

Candidate gene approaches have been frequently employed in studies of the molecular genetics of *Capsicum*. Pepper has been an ideal system for this approach due to the combination of its global crop value, its large genome size, and the ability to use comparative approaches to access the information generated from the investments in tomato as a model organism. Tomato and pepper share DNA sequence similarity, conserved synteny and the research goals of understanding and improving disease resistance and fruit size, shape, yield and quality. This chapter will explore the rationale for the candidate gene approach in pepper, case studies and the future outlook for this approach.

2. Map-Based Cloning versus the Candidate Gene Approach

Map-based cloning is a definitive method for identifying a gene underlying a phenotype. The analysis of large segregating populations allows the identification of recombination events between a phenotype and flanking genetic markers to progressively narrow and define the interval encompassing the gene of interest. This iterative process of recombination and mapping continues until only one or a few open reading frames (ORFs) remain between the segregating markers, thereby excluding the rest of the genome as the causal agent of the phenotype and linking genotype to phenotype by a process of elimination. Map-based cloning is, however, a resource-intensive process that requires the ability to genotype large populations and multiple generations to winnow away the consensus of the plant genome. Further, it depends on the occurrence of recombination events at a reasonable frequency in the vicinity of the target gene. This approach has been successfully employed multiple times to clone genes from the Solanaceae (Brommonschenkel et al. 2000; Frary et al. 2000; Chen et al. 2007; Cong et al. 2008; Orsi and Tanksley 2009) and there is at least one example of the identification of a recombination event in the target gene itself (Fridman et al. 2000). However, plant genomes contain regions with suppressed recombination, highly repetitive sequences or skewed segregation ratios, and therefore, map-based cloning of genes in these regions is much more difficult.

An alternative strategy to map-based gene cloning is the candidate gene approach, which employs a hypothesis-based method of gene identification. The hypothesis that underlies this approach is that the observed phenotype is due to a sequence polymorphism in a gene rationally associated with the trait (Pflieger et al. 2001). In addition to the ability to directly utilize hypotheses, this approach is often parsimonious and expeditious; it is reasonable to assume that a gene already known to control a trait in one organism would have the same function in a related organism. Three strategies are frequently used to create a set of candidate genes: expression, function and position.

3. Strategies for Compiling Candidate Genes

The gene expression strategy requires the least amount of *a priori* information to assemble a list of candidates. The identification of transcripts and proteins that are spatially and temporally expressed in a manner consistent with the observed phenotype is enabled by new technologies that allow large scale analysis of known genes by microarray analysis or the *de novo* sequencing of transcripts and peptides (Alba et al. 2004; Rose et al. 2004). These transcripts and proteins identify a subset of genes that may be responsible for the observed phenotype. A frequent corollary within the gene expression strategy is the use of differential expression information. Differential expression further reduces the subset of genes by applying the criterion that the expression of the target gene is enriched in the tissues and time points associated with the phenotype. Techniques which have been developed, such as suppressive subtractive hybridization (SSH), preferentially identify these transcripts (Diatchenko et al. 1996). While the gene expression strategy may not account for genes that are expressed at low levels or when a lack of expression confers the phenotype of interest, it is the most holistic approach and best suited to making novel discoveries like those found when employing map-based cloning.

Functional candidates may be developed with knowledge of the biochemical pathway or the biological process that underlies the phenotype. Structural genes are relatively straightforward to develop as candidates because their sequences allow the definition of clear homology relationships with other genes of known function and the prediction of subcellular localization. The emergence of curated pathway databases has greatly facilitated this process (Mueller et al. 2005). Heterologous sequences can be used directly as probes to find the corresponding pepper sequence by bioinformatic searches of the known pepper genes or to guide the cloning of pepper orthologs (Mazourek et al. 2009). An important but hard to predict class of genes are the regulatory genes that control these pathways and processes. The variable nature of transcription factors makes it difficult to predict structure using prior knowledge of function (Liu et al. 2003).

Genetic position candidates require a segregating population that has been phenotyped and for which molecular markers are available. Genes in the vicinity of a mapped phenotypic locus automatically become candidates provided that the sequence is available in this region. Ideally, markers will also be mapped in other species to allow the alignment of comparative maps (Livingstone et al. 1999; Wu et al. 2009; see Chapter 3). A recurring example of the comparative genetic position approach is the use of a mapped, characterized gene in tomato as a candidate for a similar phenotype in pepper. This approach introduces a second strategy of parsimony. Given the evolutionary relationship of the genera of the Solanaceae, it is parsimonious to propose that their shared biological processes are controlled by orthologous genes. Despite this underlying similarity, there is also great diversity among and between genera and species. Therefore, this strategy probes the limits of this relationship. A critical limitation of this approach is related to the availability of characterized candidates that can be tested with comparative maps.

4. Validation of Candidate Genes in *Capsicum*

In practice, combinations of the three strategies above are used to assemble and then selectively reduce a set of candidate genes. While lack of cosegregation can eliminate candidates, cosegregation only supports a candidate; further analysis with more markers or individuals may reveal recombination between the trait and the gene being tested. The inconclusive nature of the candidate gene approach compared to map-based cloning requires a greater burden of proof in the subsequent analysis. However, no further analysis is generally done because of the already thorough characterization of the orthologs of the candidate (Table 4-1).

Validating candidate genes in *Capsicum* is difficult because of the lack of a practical transformation system (Kothari et al. 2010). Transgenic complementation by the introduction of the cloned gene in a mutant pepper is the standard of proof in many organisms, but other alternatives must be sought for pepper. One approach to establish the function of a cloned pepper gene is by its transformation into another solanaceous plant with an established transformation system. Notably, this has proved to be effective for disease resistance genes transformed into tomato or *N. benthamiana*.

Virus-induced gene silencing (VIGS) can be used to validate candidate genes by knocking out gene expression without the need for transformation. It has the additional advantages of being relatively fast and capable of screening several candidate genes in parallel. VIGS utilizes the plant's natural defense against viral infection to silence the expression

Table 4-1 Candidate carotenoid genes and supporting studies.

Gene Mapped in Pepper	Phenotypic Locus	Additional study
Geranylgeranyl-pyrophosphate synthase (*Ggpps*)	*fc4.2* (tomato)	
Phytoene synthase (*Psy*)	*c2* (pepper) *rh4.1* (pepper)	*Psy* is c2 candidate (Huh et al. 2001)
Phytoene desaturase (*Pds*)	no association	
ζ-Carotene desaturase (*Zds*)	*fc1.1* (tomato)	
Lycopene ε-cyclase (*CrtL-e*)	*Del* (tomato)	(Ronen et al. 1999)
Lycopene β-cyclase (*CrtL-b*)	*l2* (tomato) *Xa* (tomato) *fc10.1* (tomato)	
β-carotene hydroxylases (*CrtZ-1, CrtZ-2*)	*rc3.1* (pepper) *rl3.1* (pepper) *Y* (potato)	β-carotene hydroxylase polymorphism cosegregates with phenotype in potato (Brown et al. 2006)
Zeaxanthin epoxidase (*Ze*)	*fc2.2* (tomato)	
Capsanthin capsorubin synthase (*Ccs*)	*pfc6.1* (pepper) *B* (tomato)	*Ccs = B* Gene cloning and phenotype of transgenic tomato (Ronen et al. 2000)
	t (tomato)	*CRTISO = Tangerine* (*t*) Map-based gene cloning and *in-vitro* activity (Isaacson et al. 2002)
	gh (tomato)	*PTOX = Ghost* (*gh*) Discovery of a loss of function mutation in a homolog of an Arabidopsis mutant (Josse et al. 2000)

of an endogenous gene. If the viral vector used for infection contains a fragment of the candidate gene, the plant's defense response will degrade the corresponding mRNA through post-transcriptional gene silencing (Baulcombe 1999). If the candidate gene is responsible for the phenotype of interest, the knock down of its expression through VIGS should result in the expected deficient phenotype.

In VIGS, the virus is introduced to the plant through agroinoculation, in which the entire viral sequence is cloned into *Agrobacterium* T-DNA for transformation (Turpen et al. 1993). Usually, a reporter gene is also cloned into the viral vector in order to assess the extent of silencing. A popular choice is phytoene desaturase, which results in the photobleaching of plant tissues when knocked down (Lu et al. 2003). Inoculation with an empty viral vector is used as a control to account for any phenotypic effect the viral infection itself may produce.

For solanaceous plants, *Tobacco rattle virus* (TRV) is the preferred viral vector for VIGS. TRV is favored because it has mild symptoms that are less likely to interfere with the phenotype of interest and because it is able to cross meristems and infect growing points (Ratcliff et al. 2001). This enables VIGS to be used to silence genes in fruits. VIGS protocols have been published which can be used in pepper (Liu et al. 2002; Chung et al. 2004).

Although VIGS can provide strong support for the identity of a candidate gene, it cannot provide definitive proof. Several caveats of VIGS result in this limitation (Lu et al. 2003). First, negative results do not rule out the gene's involvement in a phenotype. Instead, silencing may not be extensive enough to result in the expected phenotype. For this reason, it is important to measure the level of the target transcript after silencing to make sure the silencing has been effective. Second, there is the potential for non-target silencing effects, which could result in a phenotype that is not caused by silencing of the intended gene.

5. Case Studies

5.1 Disease Resistance

5.1.1 Map-based Cloning of Bs2: *Bacterial Leaf Spot Resistance*

Bacterial leaf spot caused by xanthomonads is a disease in pepper-growing regions around the world. Seed treatments and pesticides with copper, copper-containing cocktails and antibiotics have contributed to the control of the pathogen, but are readily overcome by the pathogen (Stall et al. 2009). Of all the factors conferring genetic resistance, the *Bs2* gene derived from *Capsicum chacoense* has been the most broadly deployed and durable (Cook and Guevara 1984; Kousik and Ritchie 1996; Stall et al. 2009). This durability has been explained by the loss of virulence that is observed for *Xanthomanas* strains with mutations in the corresponding pathogen gene, *avrBs2* (Kearney and Staskawicz 1990; Gassmann et al. 2000).

Given the importance of this gene, the first successful map-based cloning effort in pepper targeted the isolation of *Bs2*. A high resolution genetic map of the region was created by screening an 'Early Cal Wonder' near isogenic line for polymorphism using 300 random amplified polymorphic DNA (RAPD) primers and 319 amplified fragment length polymorphism (AFLP) double primer combinations. Promising markers were subsequently converted to sequence characterized amplified region (SCAR)/sequence tagged site (STS) markers (Tai et al. 1999a). Two markers were tightly linked to the *Bs2* locus and were present at a low copy number which made them especially appropriate for use in the cloning of *Bs2*. A yeast artificial chromosome (YAC) approach was chosen for map-based cloning of the *Bs2* gene to overcome the challenge of a large, mostly uncharacterized, genome and to span regions of repetitive DNA that would complicate the contig assembly (Tai and Staskawicz 2000). The generation of 19,000 YAC clones with an average insert size of 500 kb provided approximately 3× coverage of the pepper genome. The three markers most closely linked to *Bs2* were used to screen pooled YAC clones and two clones were identified that contained all three markers (Tai and Staskawicz 2000). A chromosome walk based on 1,800 F_2 individuals reduced the interval to 100 kb, which could be sequenced to identify six open reading frames including a nucleotide binding site (NBS)-leucine rich repeat (LLR) resistance (R) gene that was selected as the candidate for *Bs2* (Tai et al. 1999b). Coinfiltration of this NBS-LRR gene and *avrBs2* into various solanaceaous leaves elicited the hypersensitive response consistent with *Bs2*-derived resistance. Transgenic *Nicotiana benthamiana* plants and tomato plants were generated and resistance cosegregated with the transgene in secondary transformants, confirming that the cloned NBS-LRR R gene encoded *Bs2* (Tai et al. 1999b).

Despite the success of this effort, map-based gene cloning in pepper was not pursued again until recently, with the current dramatically reduced sequencing costs and new high-throughput molecular markers. The gene content of the *Bs2* YAC served to dissuade scientists from map-based gene cloning in pepper, in addition to a preponderance of retroelements in the sequenced region (Mazourek et al. 2009). The 100 kb sequenced region of the clone is bereft of genes when compared with the several genes in similarly sized NBS-LRR regions from Arabidopsis. The 27 kb intron of *Bs2* was particularly off-putting, and although this first case of map-based gene cloning in pepper appeared to be the most extreme, it was reason for many researchers to seek alternate routes of gene cloning in pepper for the next decade.

5.1.2 The L locus: Tobamovirus Resistance

Tobamoviruses are an extremely infectious group of viruses that can cause major losses in commercial pepper crops (Alonso et al. 1989). Tobamoviruses occur throughout the world and can be transmitted through infected seeds, infected soil, and through mechanical contact such as transplanting. A source of resistance to tobamoviruses was discovered in the 1980s and has been utilized in breeding since then. The *L* locus for tobamovirus resistance consists of four allelic genes—L^1, L^2, L^3 and L^4—that provide an hierarchical scope of resistance to the strains of tobamoviruses infecting pepper (Boukema 1980).

Due to the potential for high losses caused by tobamoviruses, much effort has been focused on studying the resistance the *L* locus provides. The *L* locus has been mapped to pepper chromosome 11 (Lefebvre et al. 1995, 2005) and linked markers have been generated (Lefebvre et al. 1995; Matsunaga et al. 2003; Sugita et al. 2004, 2005; Kim et al. 2008). Although these markers allow breeders to use marker-assisted selection to breed resistant plants, maximum utilization of the *L* locus requires a marker that identifies the resistance gene itself. Such a marker is additionally important because the *L* locus is tightly linked in repulsion to a major quantitative trait locus (QTL) for *Cucumber mosaic virus* (CMV) resistance (Ben Chaim et al. 2001). Cloning of the *L* locus would be informative from a pathology standpoint because the allelic relationship of the different *L* resistance alleles could be determined. Currently, it is unknown whether the *L* locus consists of alleles of one gene or consists of several tightly linked genes (Yang et al. 2009).

The study of the *L* locus is aided by the high level of synteny within the Solonaceae (Wu et al. 2009). This synteny has been observed with respect to the whole genome, but also specifically for resistance genes (Grube et al. 2000; Mazourek et al. 2009). The *L* locus maps to a region that is syntenic to a region of tomato chromosome 11 (Livingstone et al. 1999) and contains the *I2* gene for *Fusarium* resistance (Ori et al. 1997; Simons et al. 1998; Grube et al. 2000). This region is also syntenic to the region containing the *R3* locus for *Phytophthora* resistance in potato (Huang et al. 2005). Huang et al. (2005) used the synteny between tomato and potato to clone the *R3a* gene that encodes the *R3* locus using a comparative genetics approach. Based on the conservation of resistance genes previously observed (Grube et al. 2000), the *L* gene may be hypothesized to be homologous to *I2* from tomato and *R3a* from potato, but with a different pathogen specificity. Thus, *I2* and *R3a* are excellent candidate genes for *L* and have been utilized by several research programs in an attempt to further map and clone *L*.

Tomita et al. (2008) used *I2* as a candidate gene to more precisely map the *L* gene using the L^3 resistance allele in an F_2 population from a cross between a susceptible *C. annuum* and a *C. annuum* with the L^3 allele introgressed from *C. chinense*. Bulk segregant analysis was used to create two SCAR markers, one of which did not segregate from the gene and one of which was 1.5 cM from the gene. To search this region for *I2* homologs, a pepper cDNA library was screened with primers from the tomato *I2* sequence. Polymorphic, cosegregating sequences were used for fine mapping in a population of 3,391 F_2 plants. 20 recombinants identified could be used to map the *L3* gene to a region contained in five bacterial artificial chromosome (BAC) clones that could be assembled into two contigs. Stretched DNA fiber fluorescent *in situ* hybridization (FISH) analysis was used to determine the size of the gap between the contigs (about 30 kb), which indicated that

L3 had been mapped to a 350–400 kb interval of pepper chromosome 11. Southern blotting with an *I2* probe revealed many *I2* homologs, of which at least 10 resided in this interval. Tomita et al. (2008) noted that more than one gene may be involved in resistance at the *L* locus, which would make cloning more complicated.

Yang et al. (2009) used comparative genomics to create single nucleotide polymorphism (SNP) markers closely linked to the *L* gene. They also hypothesized that *L* is an ortholog of *R3* and *I2* because of the synteny of the regions. To find *I2* homologs in the pepper genome, a pepper BAC library was screened with an *I2*-derived probe. Eighty-nine positive BAC clones were detected and subsequently screened with primers from *R3* in potato. BAC clones with both *I2* and *R3a* homologs were placed on a pepper genetic map using simple sequence repeat (SSR) markers derived from the BAC sequences. BACs that mapped to the *L* locus were used to create SNP markers linked to *L*.

The mapping of a combination of known markers and these newly developed SNP markers in a single population created a comprehensive picture of the *L* locus, which was used to help determine the allelic relationship of the L^3 and L^4 resistance alleles. SNP markers found to be linked to the L^3 allele were also tested in an F_2 population segregating for L^4. For marker 189D23M, recombination between the marker and the resistance locus was significantly higher in the L^4 population compared to the L^3 population, suggesting that L^3 and L^4 may not be allelic (Yang et al. 2009).

In further work, a BAC sequence assembly of approximately 224 kb predicted 19 putative candidate genes for L^3. Six were R-gene analogs, 5 of which were homologous to *I2* and *R3a*. However, recombination between the *L* locus resistance and their markers suggests these sequences may not be candidate genes for *L*, but may be candidates for other resistance genes (Yang et al. 2009). Although current efforts have not yet resulted in the cloning of the *L* gene, the increasingly tightly linked markers that have been developed will be useful both for breeders to break undesirable linkages and in subsequent attempts to clone the *L* gene.

5.1.3 eIF4E and Potyvirus Resistance

Recessive resistance to potyviruses has been widely used in pepper breeding as a durable way to control this economically harmful group of viruses. As sources of potyvirus resistance were discovered in pepper, they were successively given unique identifiers. Kyle and Palloix (1997) condensed this list of 10 genetic symbols into five loci: *pvr1*, *pvr2*, *pvr3*, *pvr5* and *pvr6*. The durability and recessive nature of these resistances lead them to be pursued with both agricultural and scientific interests. The key to their study came from the model plant, Arabidopsis. Map-based cloning of loss-of-susceptibility mutants identified mutations in *eukaryotic translation (iso) initiation factor 4E* (*eIF(iso)4E*) as responsible for the recessive resistance to potyviruses, including *Tobacco etch virus* (TEV), an important pathogen in pepper (Lellis et al. 2002).

Ruffel et al. (2002) used this information to predict *eukaryotic translation initiation factor 4E* (*eIF4E*) as a candidate for *pvr2* in pepper. Using a tobacco clone as a probe, *eIF4E* was found to cosegregate with *Potato virus Y* (PVY) resistance in a BC_1F_1 population of 440 individuals. An *eIF4E* clone from pepper that had 73.4 percent identity to the tobacco gene was generated by rapid amplification of cDNA ends (RACE). The recessive mutant resistant allele contained mutations consistent with the predicted altered function when compared with the wild-type susceptible allele. Given the recessive nature of *pvr2*-mediated resistance, the function of *eIF4E* in susceptibility was demonstrated by transient expression of the wild-type allele in a resistant cultivar. The accumulation of PVY in resistant plants coinoculated with a construct containing the wild-type *eIF4E* allele confirmed that this gene encodes *pvr2*.

In parallel, Kang et al. (2005) used *eIF4E* as a candidate for *pvr1*, which maps to the same genomic region as *pvr2*. They amplified two pepper sequences using primers from tomato *eIF4E*. Three homologs were identified in the pepper genome, one of which cosegregated with *pvr1* on RFLP blots derived from both interspecific and intraspecific F_2 populations. Given this result, they hypothesized that *pvr1* and *pvr2* were allelic. A genetic complementation test produced F_1 progeny that was uniformly resistant to TEV. This indicated that *pvr1* and *pvr2* fail to complement (the dominant, wild-type allele encodes susceptibility) and are thus alleles of the same gene.

After discovering that *pvr1* and *pvr2* were allelic, Kang et al. (2005) sequenced the *eIF4E* coding region from peppers homozygous for each of the *pvr1* alleles, using a total of 13 genotypes. An alignment of amino acid sequences predicted from the sequence data was consistent with the resistance phenotypes; all of the resistant genotypes had several point mutations that caused non-conservative amino acid substitutions in *eIF4E*. A modeling study based on the known structure of mouse *eIF4E* indicated that the substitutions were not in highly conserved regions, but were located near functionally important amino acids.

Kang et al. (2005) went on to test their hypothesis that *eIF4E* must interact physically with the viral VPg protein for infection to occur. In both yeast and *in vitro* assays, eIF4E proteins translated from *pvr1* alleles cloned from susceptible genotypes interacted with VPg, while eIF4E from resistant genotypes did not. This confirmed their hypothesis and further validated the association between the various *pvr1* alleles and their respective resistance phenotypes.

To test if this recessive resistance gene could be used to genetically engineer resistant plants, Kang et al. (2007) overexpressed the *pvr1* allele of *eIF4E* in tomato. This overexpression resulted in dominantly inherited resistance to the same potyviruses that the allele confers resistance to in pepper. Plants expressing the *pvr1* transgene did not display infection symptoms and had enzyme-linked immunosorbent assay (ELISA) values similar to non-inoculated plants. They hypothesized that dominant resistance is achieved through eIF4E's mode of action. It is predicted that eIF4E is required to bind to another protein, eIF4G, in addition to the viral protein VPg, in order to permit viral infection. Kang et al. predicted that the overexpression of ectopic "resistant eIF4E" may cause it to bind to eIF4G in the tomato plants, saturating eIF4G with the "resistant eIF4E" and consequently preventing eIF4G's interaction with the virus. This system would explain the dominant resistance caused by ectopic *pvr1* expression. This hypothesis was supported by yeast-2-hybrid experiments which showed that all of the *Capsicum* eIF4E proteins associated with resistance retain the ability to interact with Arabidopsis eIF4G.

In a more detailed analysis of the amino acid substitutions in the resistant eIF4E proteins, Yeam et al. (2007) created six new *eIF4E* alleles, each with only one of the amino acid substitutions from the resistant alleles. These six alleles were first tested for their ability to bind to VPg using a yeast-2-hybrid assay, because previous studies indicated that eIF4E proteins that confer resistance should not interact with VPg. Two amino acid substitutions found in both *pvr1¹* and *pvr1²* caused eIF4E to bind inconsistently with VPg. The authors predicted that both of the substitutions are necessary to prevent binding. Another amino acid substitution (G107R), found in the *pvr1* allele, resulted in a loss of Vpg binding. This substitution was chosen for further study.

The lack of interaction with VPg caused by this substitution was confirmed *in planta* through an *Agrobacterium* transient expression assay in tobacco. Two constructs were created in which one half of yellow fluorescence protein (YFP) was fused to the engineered eIF4E and the other half was fused to VPg. If the two proteins interacted *in planta*, the two YFP halves should have been brought together, resulting in detectable yellow fluorescence. Expression of the susceptible *pvr1⁺* allele resulted in fluorescence as expected, but expression of the *eIF4E* allele with the G107R substitution did not display fluorescence, indicating that the G107R substitution prevents the interaction of eIF4E and VPg *in planta*.

Yeam et al. (2007) also used overexpression in tomato to test whether the G107R amino acid substitution was sufficient to confer dominant resistance. They transformed tomato with the engineered *eIF4E* allele that had only the G107R substitution. They found that this substitution gave complete resistance, even to one virus strain to which the *pvr1* allele does not provide complete resistance.

The study of the *pvr* loci and *eIF4E* benefited greatly from the use of the candidate gene approach. *eIF4E* was chosen as a candidate based on the current literature and its cloning was aided by the use of tobacco and tomato *eIF4E* sequences. In addition, information about the role of the eIF4E protein in other species allowed the researchers to hypothesize about the functional role of eIF4E and the way it provides resistance. They predicted the effect of amino acid substitutions using known eIF4E structures from other species. Finally, the knowledge gained in pepper is likely to to apply to other important plant species and may permit the creation of resistant plants through transgenics.

5.2 Flavor

Pungency caused by the presence of capsaicinoids in the fruit is perhaps the most notable characteristic of the genus *Capsicum*. It is this quality that drives the use of pepper as a spice and lead to its global dispersal from the New World. A mutation that abolishes the accumulation of capsaicinoids in fruit has been known for hundreds of years and was described as a single recessive locus. This locus, *pun1* (formerly *c*), was finally elucidated in 2005 after a lengthy process of screening candidate genes (Blum et al. 2003; Stewart et al. 2005). Curry et al. (1999) described a strategy for isolating genes related to the capsaicinoid biosynthetic pathway. Capsaicinoids are synthesized in the placental dissepiment during a finite period early in fruit development. Therefore, transcripts that were differentially expressed in that time and place could be candidates for capsaicinoid biosynthesis genes and thus candidates for *pun1*.

Differential expression was used as a tool to generate several such candidates (Aluru et al. 2003). Most were assumed to be active in capsaicinoid biosynthesis because they bore sequence similarity to enzymes that fit the capsaicinoid biosynthetic model. Further study on a subset of these candidates has so far consistently supported that assumption (Aluru et al. 2003; Abraham-Juarez et al. 2008). A restriction fragment length polymorphism (RFLP)-based mapping strategy was employed that systematically eliminated these genes as encoding *Pun1* (Blum et al. 2003).

An SSH library that was developed by subtracting leaf transcripts from ripe habanero fruit transcripts produced an intriguing clone. Although the library was derived from a time point that was too late in fruit development to support capsaicinoid biosynthesis, the first clone from this library that was placed on a genetic map, *1a2*, colocalized with *Pun1* (Blum et al. 2002).

Further analysis, however, eliminated *1a2* as a candidate for *pun1* when recombination was discovered between a polymorphism in *1a2* and *pun1*, which highlights the potential for false positives when using candidate genes. Another study that developed an SSH library to obtain transcripts unique to placental tissue identified a gene fragment, SB2-66 (Kim et al. 2001), that colocalized with *Pun1* in an interspecific mapping population of 242 individuals derived from a cross between a pungent pepper and one known to carry the *pun1* allele (Stewart et al. 2005). The full sequence of SB2-66 predicted a gene that was the third acyltransferase found to be expressed in pepper fruit, and so renamed *AT3*. Further study of a panel of peppers indicated the loss of a band on RFLP blots consistent with the later sequencing of the mutant allele that identified a 2.5kb deletion that spanned the putative promoter region as well as much of the first exon of *AT3*. Expression of *AT3* was found to be consistent with the differential expression model for the pungency-related genes; *AT3* was expressed at the onset of capsaicin biosynthesis and restricted to the capsaicinoid synthesizing tissue. Functionally, *AT3* was verified as *Pun1* by VIGS with TRV. TRV infection alone increased capsaicinoid accumulation, while silencing with TRV::*AT3* resulted in pungency levels lower than either uninoculated plants or plants infected with TRV alone.

While this experiment confirmed *AT3* as *Pun1*, the biochemical role of *AT3* remains elusive. Based on homology, *AT3* is a member of the BAHD superfamily of acyltransferases that catalyze diverse acylation reactions related to plant specialized metabolism, a gene family predicted to include capsaicin synthase, the final enzyme of capsaicinoid biosynthesis that acylates vanillylamine with isodecenoyl-CoA. Numerous attempts to demonstrate this enzyme activity for *AT3* have failed because of either the insolubility of the recombinant protein (Stewart et al. 2005) or a loss of activity of the native protein past initial purification steps (M. Mazourek, unpublished data). It has, therefore, become tempting to speculate about *AT3* as a regulator of capsaicin biosynthesis. Evidence supporting this role is found in the loss of the transcription of capsaicinoid biosynthetic genes in a *pun1* background and the demonstration that a member of this protein family is imported to the nucleus (Yu et al. 2008).

Loss of pungency mutants have also been reported in species outside of *C. annuum*. *Pun1* was an obvious candidate for these mutants in *C. chinense*, *C. frutescens* and *C. chacoense*. In *C. chinense* and *C. frutescens*, loss of pungency cosegregated with *Pun1* and a mutation consistent with loss of function of the gene was found. In *C. chinense* (*pun1²*), a -4 bp frameshift mutation in the first exon disrupts the protein sequence and introduces a premature stop codon (Stewart et al. 2007). In *C. frutescens* (*pun1³*), truncation of the second exon eliminates a conserved motif (Stellari et al. 2010). In both cases complementation tests with *pun1¹* (*C. annuum*) unequivocally identified the mutants as allelic. *C. chacoense* is a species outside the closely related *annuum-chinense-frutescens* complex and, accordingly, the loss of pungency locus in *C. chacoense* is a separate gene, *pun2*, as demonstrated by mapping and complementation (Stellari et al. 2010).

5.3 Colors

5.3.1 A *and* cl: *Purple and Chocolate Coloration*

Anthocyanin biosynthesis derives from the phenylpropanoid pathway and is regulated by a suite of transcription factors. Despite the complexity of factors involved in anthocyanin biosynthesis, inheritance of this pigmentation is controlled by relatively few Mendelian characters. Comparative maps of the Solanaceae showed that genes controlling the presence of anthocyanins in eggplant, potato and pepper colocalized to the same genomic region (Doganlar et al. 2002; Ben Chaim et al. 2003; De Jong et al. 2004). Following the finding that the petunia transcription factor *An2* mapped to the corresponding position in potato (De Jong et al. 2004), Borovsky and colleagues hypothesized that a pepper ortholog of *An2, A*, was responsible for anthocyanin accumulation (Borovsky et al. 2004). RFLP blots of F_2 populations that segregated for the anthocyanin accumulation were probed with the petunia *An2* gene and revealed a single homolog in pepper that further cosegregated with anthocyanin accumulation in all 295 individuals. A fragment of this locus was amplified with PCR primers designed to a conserved region of *An2* and the complete sequence was obtained by RACE. Expression analysis of *A* from pepper showed that transcript abundance was consistent with phenotype, with expression throughout anthocyanin-accumulating plants except for ripe fruits. Plants that did not accumulate anthocyanins had no detectable transcripts. Further, structural genes of anthocyanin biosynthesis were expressed in a similar fashion to *A*, which supports the role of the candidate as a transcription factor that influences their expression.

Chocolate peppers are the result of a failure of chlorophyll breakdown during ripening because of a single recessive gene, *cl*. The mixing of chlorophyll with newly formed red carotenoids gives peppers a brown (chocolate) appearance, or in peppers that fail to form carotenoids, the immature green color will be maintained through ripening. *Stay-green* mutants, defective in pheophorbide *a* oxidase activity, are known in other species and similarities between the chlorophyll metabolites of brown-fruited peppers and *stay-green* mutants lead to the hypothesis that the *cl* locus in pepper is encoded by a homolog of this characterized gene, *pheophorbide a oxidase* (*PaO*) (Roca and Minguez-Mosquera 2006; Borovsky and Paran 2008). Using a BC_1F_1 population of 198 individuals from an interspecific cross between a red-fruited

C. chinense and a green-fruited *C. annuum* as the recurrent parent, a pepper stay-green homolog (*CaSGR*) was shown to cosegregate with the chlorophyll-retaining phenotype. The expression of *CaSGR* agreed with this mapping result; *CaSGR* mRNA expression was higher in red fruits that degraded chlorophyll than in fruit that did not degrade chlorophyll (Borovsky and Paran 2008). Further, a missense mutation was detected within the mutant allele. Barry et al. (2008) were able to use the map position of *cl* in pepper to aid the positional cloning in tomato of the *green-flesh* locus, *gf*. They also cloned *cl* from pepper and found the same mutation as did Borovsky and Paran, but did not find the difference in the gene expression between mutant and wild-type fruits. Transgenic complementation of the *gf* mutation with the wild-type tomato allele strongly supports the proposed role of the pepper ortholog *CaSGR* as *cl*.

5.3.2 Carotenoid Variation in Mature Fruit

Mature fruit color of pepper is determined primarily by carotenoid content (see chapter 1 and 2). Huh et al. (2001) and Thorup et al. (2000) pursued the *c2* gene of pepper that, when in the homozygous recessive state, is responsible for an orange rather than red mature fruit color. They both used a panel of cloned structural genes from the carotenoid biosynthetic pathway as candidates for the locus. Thorup focused on a BC_1F_1 population derived from a cross between a red fruited *C. chinense* and a white fruited *C. annuum* and identified four phenotypic classes for fruit color that were attributed to the action of two genes, *c1* and *c2*. Huh et al. (2001) also used an interspecific pepper population derived from a cross between a red fruited *C. annuum* and an orange fruited *C. chinense*. Among 103 F_2 individuals, phytoene synthase, an early gene in the carotenoid biosynthetic pathway, showed complete linkage between fruit color and a polymorphism on RFLP blots. A difference in carotenoid abundance, rather than variation in carotenoid profile, was determined to be the reason behind contrasting fruit colors. Thorup et al. (2000) further investigated the use of candidate genes for fruit pigmentation in the Solanaceae by extending this analysis to a comparative system. In addition to some classical qualitative loci affecting pepper color, they sought to explore the range of carotenoid variation in the Solanceae by looking outward from pepper using a comparative strategy. Ten structural genes that encode enzymes for carotenoid biosynthesis were mapped in a pepper population that was linked with the tomato and potato genomes through a comparative map. While phytoene desaturase and the specific homolog of β-carotene hydroxylase could not be linked with a phenotype, the remainder of the genes tested aligned with the previously described loci in tomato, pepper and/or potato (Table 4-1).

Contemporaneous activities focused on individual major effect genes, which generally support the functions inferred from this colocalization on comparative maps. However, as noted by Liu et al. (2003), the role of these genes as candidates for quantitative variation in carotenoid accumulation has not been supported. Also, due to the focus on structural genes, transcription factors related to the fruit development were not included as candidates in the analysis.

6. Future Outlook

The candidate gene approach has been applied with much success in pepper. Largely this success has been because of the ability to work within the Solanaceae as a highly developed comparative genetic system. The candidate gene approach has served two purposes. First, it has served to test hypotheses framed around the conservation of gene function in the Solanaceae. Second, it has been used to efficiently characterize traits of agricultural importance in pepper despite the daunting size of the pepper genome (see chapter 6).

The candidate gene approach has not been as fruitful in human genetics, where critics have noted its failure to deliver based on "shot in the dark" predictions (Altshuler et al. 2008). This concern about the ability of the candidate gene approach to reveal novel or complex discoveries has been echoed in the plant community, where genome wide studies are becoming much more common (Myles et al. 2009). One has to question the fate of the candidate gene approach in the post-genomic era. We are optimistic about its continued utility for the following reasons:

First, the candidate gene approach remains the most efficient means to an end in many cases. Rather than dismiss the investments into the comparative system of the Solanaceae and the investment in tomato as a model organism, it is logical to use these resources to their fullest. Many questions of agronomic importance remain in pepper and the candidate gene approach continues to be effective and less resource intensive than the more modern method of association analysis with high density marker coverage.

Second, modern population analysis methodology allows the rapid identification of a genomic region of interest. Next generation sequence technology readily discovers potential open reading frames within this region. In all cases where multiple open reading frames are identified, these putative genes need to be prioritized for analysis, often using a candidate gene approach (Ni et al. 2009). It is our opinion that we will increasingly move toward this modified version of the candidate gene approach in the future.

Third, the candidate gene approach probes questions about the conservation of gene function and evolution within the Solanaceae and can reveal surprising examples of divergent evolution. The red pigments capsanthin and capsorubin, found in most ripe pepper fruits, are relatively unique in nature and one of the distinctive contrasts between pepper and tomato. Ripe fruit color in each crop is influenced by orthologs of a lycopene β-cyclase. In tomato, an allele of this gene allows the accumulation of the orange pigment β-carotene at the expense of the red pigment lycopene. In pepper, the orthologous enzyme is required for the synthesis of the red pigments capsanthin and capsorubin from yellow xanthophylls. A comparative genetics study of fruit color that used candidate genes cloned from the carotenoid biosynthetic pathway (Thorup et al. 2000) revealed this association in parallel to a conventional forward genetics study (Ronen et al. 2000). Such questions about the genomic changes that underlie speciation may be best addressed through candidate gene approaches that incorporate comparative genomic hypotheses.

References

Abraham-Juarez MD, Rocha-Granados MD, Lopez MG, Rivera-Bustamante RF, Ochoa-Alejo N (2008) Virus-induced silencing of *Comt*, *pAmt* and *Kas* genes results in a reduction of capsaicinoid accumulation in chili pepper fruits. Planta 227: 681–695.

Alba R, Fei Z, Payton P, Liu Y, Moore SL, Debbie P, Cohn J, D'Ascenzo M, Gordon JS, Rose JKC, Martin G, Tanksley SD, Bouzayen M, Jahn MM, Giovannoni J (2004) ESTs, cDNA microarrays, and gene expression profiling: tools for dissecting plant physiology and development. Plant J 39: 697–714.

Alonso E, Garcia-Luque I, Avila-Rincon MJ, Wichh B, Serra MT, Diazruiz JR (1989) A tobamovirus causing heavy losses in protected pepper crops in Spain. J Phytopathol 125: 67–76.

Altshuler D, Daly MJ, Lander ES (2008) Genetic mapping in human disease. Science 322: 881–888.

Aluru MR, Mazourek M, Landry LG, Curry J, Jahn M, O'Connell MA (2003) Differential expression of fatty acid synthase genes, *Acl*, *Fat* and *Kas*, in *Capsicum* fruit. J Exp Bot 54: 1655–1664.

Barry CS, McQuinn RP, Chung MY, Besuden A, Giovannoni JJ (2008) Amino acid substitutions in homologs of the STAY-GREEN protein are responsible for the *green-flesh* and *chlorophyll retainer* mutations of tomato and pepper. Plant Physiol 147: 179–187.

Baulcombe DC (1999) Fast forward genetics based on virus-induced gene silencing. Curr Opin Plant Biol 2: 109–113.

Ben Chaim A, Borovsky Y, De Jong W, Paran I (2003) Linkage of the *A* locus for the presence of anthocyanin and *fs10.1*, a major fruit-shape QTL in pepper. Theor Appl Genet 106: 889–894.

Ben Chaim A, Grube RC, Lapidot M, Jahn M, Paran I (2001) Identification of quantitative trait loci associated with resistance to cucumber mosaic virus in *Capsicum annuum*. Theor Appl Genet 102: 1213–1220.

Blum E, Liu K, Mazourek M, Yoo EY, Jahn M, Paran I (2002) Molecular mapping of the *C* locus for presence of pungency in *Capsicum*. Genome 45: 702–705.

Blum E, Mazourek M, O'Connell M, Curry J, Thorup T, Liu KD, Jahn M, Paran I (2003) Molecular mapping of capsaicinoid biosynthesis genes and quantitative trait loci analysis for capsaicinoid content in *Capsicum*. Theor Appl Genet 108: 79–86.

Borovsky Y, Oren-Shamir M, Ovadia R, De Jong W, Paran I (2004) The *A* locus that controls anthocyanin accumulation in pepper encodes a MYB transcription factor homologous to *Anthocyanin2* of Petunia. Theor Appl Genet 109: 23–29.

Borovsky Y, Paran I (2008) Chlorophyll breakdown during pepper fruit ripening in the *chlorophyll retainer* mutation is impaired at the homolog of the senescence-inducible stay-green gene. Theor Appl Genet 117: 235–240.

Boukema IW (1980) Allelism of Genes-Controlling Resistance to TMV in *Capsicum* L. Euphytica 29: 433–439.

Brommonschenkel SH, Frary A, Frary A, Tanksley SD (2000) The broad-spectrum tospovirus resistance gene *Sw-5* of tomato is a homolog of the root-knot nematode resistance gene *Mi*. Mol Plant Microbe Interact 13: 1130–1138.

Brown CR, Kim TS, Ganga Z, Haynes K, De Jong D, Jahn M, Paran I, De Jong W (2006) Segregation of total carotenoid in high level potato germplasm and its relationship to *beta-carotene hydroxylase* polymorphism. Amer J Potato Res 83: 365–372.

Chen KY, Cong B, Wing R, Vrebalov J, Tanksley SD (2007) Changes in regulation of a transcription factor lead to autogamy in cultivated tomatoes. Science 318: 643–645.

Chung E, Seong E, KimYC, Chung EJ, Oh SK, Lee S, Park JM, Joung YH, Choi D (2004) A method of high frequency virus-induced gene silencing in chili pepper (*Capsicum annuum* L. cv. Bukang). Mol Cells 17: 377–380.

Cong B, Barrero LS, Tanksley SD (2008) Regulatory change in YABBY-like transcription factor led to evolution of extreme fruit size during tomato domestication. Nat Genet 40: 800–804.

Cook AA, Guevara YG (1984) Hypersensitivity in *Capsicum chacoense* to Race-1 of the bacterial spot pathogen of pepper. Plant Dis 68: 329–330.

Curry J, Aluru M, Mendoza M, Nevarez J, Melendrez M, O'Connell MA (1999) Transcripts for possible capsaicinoid biosynthetic genes are differentially accumulated in pungent and non-pungent *Capsicum* spp. Plant Sci 148: 47–57.

De Jong WS, Eannetta NT, De Jong DM, Bodis M (2004) Candidate gene analysis of anthocyanin pigmentation loci in the Solanaceae. Theor Appl Genet 108: 423–432.

Diatchenko L, Lau YFC, Campbell AP, Chenchik A, Moqadam F, Huang B, Lukyanov S, Lukyanov K, Gurskaya N, Sverdlov ED, Siebert PD (1996) Suppression subtractive hybridization: A method for generating differentially regulated or tissue-specific cDNA probes and libraries. Proc Natl Acad Sci USA 93: 6025–6030.

Doganlar S, Frary A, Daunay MC, Lester RN, Tanksley SD (2002) Conservation of gene function in the Solanaceae as revealed by comparative mapping of domestication traits in eggplant. Genetics 161: 1713–1726.

Frary A, Nesbitt TC, Frary A, Grandillo S, van der Knaap E, Cong B, Liu JP, Meller J, Elber R, Alpert KB, Tanksley SD (2000) *fw2.2*: A quantitative trait locus key to the evolution of tomato fruit size. Science 289: 85–88.

Fridman E, Pleban T, Zamir D (2000) A recombination hotspot delimits a wild-species quantitative trait locus for tomato sugar content to 484 bp within an invertase gene. Proc Natl Acad Sci USA 97: 4718–4723.

Gassmann W, Dahlbeck D, Chesnokova O, Minsavage GV, Jones JB, Staskawicz BJ (2000) Molecular evolution of virulence in natural field strains of *Xanthomonas campestris* pv. *vesicatoria*. J Bacteriol 182: 7053–7059.

Grube RC, Radwanski ER, Jahn M (2000) Comparative genetics of disease resistance within the solanaceae. Genetics 155: 873–887.

Huang SW, van der Vossen EAG, Kuang HH, Vleeshouwers V, Zhang NW, Borm TJA, van Eck HJ, Baker B, Jacobsen E, Visser RGF (2005) Comparative genomics enabled the isolation of the *R3a* late blight resistance gene in potato. Plant J 42: 251–261.

Huh JH, Kang BC, Nahm SH, Kim S, Ha KS, Lee MH, Kim BD (2001) A candidate gene approach identified phytoene synthase as the locus for mature fruit color in red pepper (*Capsicum* spp.). Theor Appl Genet 102: 524–530.

Isaacson T, Ronen G, Zamir D, Hirschberg J (2002) Cloning of *tangerine* from tomato reveals a carotenoid isomerase essential for the production of beta-carotene and xanthophylls in plants. Plant Cell 14: 333–342.

Josse EM, Simkin AJ, Gaffe J, Laboure AM, Kuntz M, Carol P (2000) A plastid terminal oxidase associated with carotenoid desaturation during chromoplast differentiation. Plant Physiol 123: 1427–1436.

Kang BC, Yeam I, Frantz JD, Murphy JF, Jahn MM (2005) The *pvr1* locus in *Capsicum* encodes a translation initiation factor eIF4E that interacts with Tobacco etch virus VPg. Plant J 42: 392–405.

Kang BC, Yeam I, Li HX, Perez KW, Jahn MM (2007) Ectopic expression of a recessive resistance gene generates dominant potyvirus resistance in plants. Plant Biotechnol J 5: 526–536.

Kearney B, Staskawicz BJ (1990) Widespread distribution and fitness contribution of *Xanthomonas campestris* Avirulence gene *Avrbs2*. Nature 346: 385–386.

Kim HJ, Han JH, Yoo JH, Cho HJ, Kim BD (2008) Development of a sequence characteristic amplified region marker linked to the L^4 locus conferring broad spectrum resistance to tobamoviruses in pepper plants. Mol Cells 25: 205–210.

Kim M, Kim S, Kim S, Kim BD (2001) Isolation of cDNA clones differentially accumulated in the placenta of pungent pepper by suppression subtractive hybridization. Mol Cells 11: 213–219.

Kothari SL, Joshi A, Kachhwaha S, Ochoa-Alejo N (2010) Chilli peppers—A review on tissue culture and transgenesis. Biotechnol Adv 28: 35–48.

Kousik CS, Ritchie DF (1996) Disease potential of pepper bacterial spot pathogen races that overcome the *Bs2* gene for resistance. Phytopathology 86: 1336–1343.

Kyle MM, Palloix A (1997) Proposed revision of nomenclature for potyvirus resistance genes in *Capsicum*. Euphytica 97: 183–188.

Lefebvre V (2005) Molecular markers for genetics and breedings: Development and use in pepper (*Capsicum* spp.). In: Lorz H, Wenzel G [ed] Biotechnology in agriculture and forestry, vol 55. Molecular marker systems in plant breeding and crop improvement, Springer, Berlin Heidelberg, New York, USA pp 189–214.

Lefebvre V, Palloix A, Caranta C, Pochard E (1995) Construction of an intraspecific integrated linkage map of pepper using molecular markers and doubled haploid progenies. Genome 38: 112–121.

Lellis AD, Kasschau KD, Whitham SA, Carrington JC (2002) Loss-of-susceptibility mutants of Arabidopsis thaliana reveal an essential role for eIF(iso)4E during potyvirus infection. Curr Biol 12: 1046–1051.

Liu YL, Schiff M, Marathe R, Dinesh-Kumar SP (2002) Virus-induced gene silencing in tomato. Plant J 31: 777–786.

Liu YS, Gur A, Ronen G, Causse M, Damidaux R, Buret M, Hirschberg J, Zamir D (2003) There is more to tomato fruit colour than candidate carotenoid genes. Plant Biotechnol J 1: 195–207.

Livingstone KD, Lackney VK, Blauth JR, van Wijk R, Jahn MK (1999) Genome mapping in *Capsicum* and the evolution of genome structure in the *Solanaceae*. Genetics 152: 1183–1202.

Lu R, Martin-Hernandez AM, Peart JR, Malcuit I, Baulcombe DC (2003) Virus-induced gene silencing in plants. Methods 30: 296–303.

Matsunaga H, Saito T, Hirai M, Nunome T, Yoshida T (2003) DNA markers linked to pepper mild mottle virus (PMMoV) resistant locus (L^4) in *Capsicum*. J Jpn Soc Hort Sci 72: 218–220.

Mazourek M, Cirulli ET, Collier SM, Landry LG, Kang BC, Quirin EA, Bradeen JM, Moffett P, Jahn MM (2009) The fractionated orthology of *Bs2* and *Rx/Gpa2* supports shared synteny of disease resistance in the Solanaceae. Genetics 182: 1351–1364.

Mueller LA, Solow TH, Taylor N, Skwarecki B, Buels R, Binns J, Lin CW, Wright MH, Ahrens R, Wang Y, Herbst EV, Keyder ER, Menda N, Zamir D, Tanksley SD (2005) The SOL Genomics Network: a comparative resource for Solanaceae biology and beyond. Plant Physiol 138: 1310–1317.

Myles S, Peiffer J, Brown PJ, Ersoz ES, Zhang ZW, Costich DE, Buckler ES (2009) Association mapping: Critical considerations shift from genotyping to experimental design. Plant Cell 21: 2194–2202.

Ni J, Pujar A, Youens-Clark K, Yap I, Jaiswal P, Tecle I, Tung C-W, Ren L, Spooner W, Wei X, Avraham S, Ware D, Stein L, McCouch S (2009) Gramene QTL database: development, content and applications. Database(Oxford). 2009: bap005.

Ori N, Eshed Y, Paran I, Presting G, Aviv D, Tanksley S, Zamir D, Fluhr R (1997) The *I2C* family from the wilt disease resistance locus *I2* belongs to the nucleotide binding, leucine-rich repeat superfamily of plant resistance genes. Plant Cell 9: 521–532.

Orsi CH, Tanksley SD (2009) Natural variation in an ABC transporter gene associated with seed size evolution in tomato species. PLoS Genet. 5(1):e 1000347.

Pflieger S, Lefebvre V, Causse M (2001) The candidate gene approach in plant genetics: a review. Mol Breed 7: 275–291.

Ratcliff F, Martin-Hernandez AM, Baulcombe DC (2001) Tobacco rattle virus as a vector for analysis of gene function by silencing. Plant J 25: 237–245.

Roca M, Minguez-Mosquera MI (2006) Chlorophyll catabolism pathway in fruits of *Capsicum annuum* (L.): Stay-green versus red fruits. J Agri Food Chem 54: 4035–4040.

Ronen G, Carmel-Goren L, Zamir D, Hirschberg J (2000) An alternative pathway to beta-carotene formation in plant chromoplasts discovered by map-based cloning of *Beta* and *old-gold* color mutations in tomato. Proc Natl Acad Sci USA 97: 11102–11107.

Ronen G, Cohen M, Zamir D, Hirschberg J (1999) Regulation of carotenoid biosynthesis during tomato fruit development: expression of the gene for lycopene epsilon-cyclase is down-regulated during ripening and is elevated in the mutant Delta. Plant J 17: 341–351.

Rose JK, Bashir S, Giovannoni JJ, Jahn MM, Saravanan RS (2004) Tackling the plant proteome: practical approaches, hurdles and experimental tools. Plant J 39: 715–733.

Ruffel S, Dussault MH, Palloix A, Moury B, Bendahmane A, Robaglia C, Caranta C (2002) A natural recessive resistance gene against potato virus Y in pepper corresponds to the *eukaryotic initiation factor 4E* (*eIF4E*). Plant J 32: 1067–1075.

Simons G, Groenendijk J, Wijbrandi J, Reijans M, Groenen J, Diergaarde P, Van der Lee T, Bleeker M, Onstenk J, de Both M, Haring M, Mes J, Cornelissen B, Zabeau M, Vos P (1998) Dissection of the Fusarium *I2* gene cluster in tomato reveals six homologs and one active gene copy. Plant Cell 10: 1055–1068.

Stall RE, Jones JB, Minsavage GV (2009) Durability of resistance in tomato and pepper to Xanthomonads causing bacterial spot. Annu Rev Phytopathol 47: 265–284.

Stellari GM, Mazourek M, Jahn MM (2010) Contrasting modes for loss of pungency between cultivated and wild species of *Capsicum*. Heredity 104: 460–471.

Stewart C, Kang BC, Liu K, Mazourek M, Moore SL, Yoo EY, Kim BD, Paran I, Jahn MM (2005) The *Pun1* gene for pungency in pepper encodes a putative acyltransferase. Plant J 42: 675–688.

Stewart C, Mazourek M, Stellari GM, O'Connell M, Jahn M (2007) Genetic control of pungency in *C. chinense* via the *Pun1* locus. J Exp Bot 58: 979–991.

Sugita T, Kinoshita T, Kawano T, Yuji K, Yamaguchi K, Nagata R, Shimizu A, Chen LZ, Kawasaki S, Todoroki A (2005) Rapid construction of a linkage map using high-efficiency genome scanning/AFLP and RAPD, based on an intraspecific, doubled-haploid population of *Capsicum annuum*. Breed Sci 55: 287–295.

Sugita T, Yamaguchi, Sugimura Y, Nagata R, Yuji K, Kinoshita T, Todoroki A (2004) Development of SCAR markers linked to *L3* gene in *Capsicum*. Breed Sci 54: 111–115.

Tai T, Dahlbeck D, Stall RE, Peleman J, Staskawicz BJ (1999a) High-resolution genetic and physical mapping of the region containing the *Bs2* resistance gene of pepper. Theor Appl Genet 99: 1201–1206.

Tai T, Staskawicz BJ (2000) Construction of a yeast artificial chromosome library of pepper (*Capsicum annuum* L.) and identification of clones from the *Bs2* resistance locus. Theor Appl Genet 100: 112–117.

Tai TH, Dahlbeck D, Clark ET, Gajiwala P, Pasion R, Whalen MC, Stall RE, Staskawicz BJ (1999b) Expression of the *Bs2* pepper gene confers resistance to bacterial spot disease in tomato. Proc Natl Acad Sci U S A 96: 14153–14158.

Thorup TA, Tanyolac B, Livingstone KD, Popovsky S, Paran I, Jahn M 2000 Candidate gene analysis of organ pigmentation loci in the *Solanaceae*. Proc Natl Acad Sci U S A 97: 11192–11197.

Tomita R, Murai J, Miura Y, Ishihara H, Liu S, Kubotera Y, Honda A, Hatta R, Kuroda T, Hamada H, Sakamoto M, Munemura I, Nunomura O, Ishikawa K, Genda Y, Kawasaki S, Suzuki K, Meksem K, Kobayashi K (2008) Fine mapping and DNA fiber FISH analysis locates the tobamovirus resistance gene *L3* of *Capsicum chinense* in a 400-kb region of R-like genes cluster embedded in highly repetitive sequences. Theor Appl Genet 117: 1107–1118.

Turpen TH, Turpen AM, Weinzettl N, Kumagai MH, Dawson WO (1993) Transfection of whole plants from wounds inoculated with *Agrobacterium Tumefaciens* containing Cdna of *Tobacco Mosaic Virus*. J Virol Methods 42: 227–240.

Wu FN, Eannetta NT, Xu YM, Durrett R, Mazourek M, Jahn, Tanksley SD (2009) A COSII genetic map of the pepper genome provides a detailed picture of synteny with tomato and new insights into recent chromosome evolution in the genus *Capsicum*. Theor Appl Genet 118: 1279–1293.

Yang HB, Liu WY, Kang WH, Jahn M, Kang BC (2009) Development of SNP markers linked to the *L* locus in *Capsicum* spp. by a comparative genetic analysis. Mol Breed 24: 433–446.

Yeam I, Cavatorta JR, Ripoll DR, Kang BC, Jahn MM (2007) Functional dissection of naturally occurring amino acid substitutions in eIF4E that confers recessive potyvirus resistance in plants. Plant Cell 19: 2913–2928.

Yu XH, Chen MH, Liu CJ (2008) Nucleocytoplasmic-localized acyltransferases catalyze the malonylation of 7-O-glycosidic (iso)flavones in *Medicago truncatula*. Plant J 55: 382–396.

5

Molecular Mapping of Complex Traits in *Capsicum*

James P. Prince,[1,]* *Davis Cheng*[1] and *Cristina Fernández Otero*[2]

ABSTRACT

In this chapter, we will examine the basics of mapping the location of genes controlling complex traits in pepper (*Capsicum*). First, we will briefly examine variation in the genus. Next we will discuss what a complex trait is (and define QTL or quantitative trait locus), as well as provide an overview of how a finite number of genes can lead to the continuous level of variation common in many quantitative traits. Then we will provide a list of complex traits for which QTLs have been mapped in the genus. The bulk of the chapter will show how QTLs are analyzed, including phenotypic issues, an introduction to the main types of DNA markers that have been used to map QTLs in pepper, a discussion on the different types of mapping populations that have been used, a comparison of single-marker analysis, interval analysis, and composite interval analysis and a listing of QTL analysis software commonly used in pepper. The chapter ends with comments on progress in the Mendelization of pepper QTLs.

Keywords: Quantitative trait loci (QTL), Molecular marker, Population, Marker-assisted selection (MAS), Mapping

1. Variation in *Capsicum*

Capsicum spp., the sweet pepper or chili pepper genus, is full of variation in fruit color, size, shape, pungency and other flavour characteristics, as

[1]Department of Biology, California State University Fresno, Fresno CA 93740 USA.
[2]Centro de Investigaciones Agrarias de Mabegondo, A Coruña, Spain.
*Corresponding author

well as, to a lesser extent, variation in flower and leaf form (Chapter 1 and Andrews 1984). In addition, there is considerable variation in levels of disease resistance. Variation at the molecular level in pepper is somewhat more complicated and harder to quantify. Nevertheless, the variation seen provides for extensive opportunities for genetic analyses as well as for esthetic pleasure. Who can resist exploring the gastronomic delights provided by the variety of peppers available to most people today or fail to be delighted by the array of sizes and shapes from the tiny bird peppers to long cayenne types to the giant boxy bells and colors from dark green to purple to "black" to orange, red, and yellow (Chapter 1)? In the United States, hot peppers have been growing in culinary popularity for decades. There is also a large group of people who enjoy the endorphin "buzz" provided by recovering from eating the very hot peppers such as Habanero or the Ghost Pepper.

What controls the phenotypic variation seen in *Capsicum*? Many single genes controlling phenotypes of interest have been indentified, and some have been mapped onto the genome of pepper (Chapter 4). Most traits, however, are complex and have only been accessible to analysis and mapping in the last 20 years. New technologies and high-throughput analyses have increased our ability to make sense of the genes controlling phenotypes of horticultural interest.

2. What is a Complex Trait? What are Quantitative Trait Loci (QTL)?

A genetically complex trait (quantitative trait) is one whose patterns of inheritance cannot be explained readily by any single-gene model. The phenotypic distribution is typically continuous and it is often a Normal distribution. Usually the trait is one encoded for by multiple genes and it often has a heritability of less than 10 percent (indicating noticeable environmental effects on the phenotype). Such a trait is also often called a polygenic trait. Examples of such traits in pepper are yield, fruit color, germination rate, fruit size and shape and many disease resistance phenotypes.

More formally, the variation of phenotype of any trait is represented by the total variance, σ_T^2, which can be broken into the following components:

$$\sigma_T^2 = \sigma_G^2 + \sigma_E^2 + \sigma_{GxE}^2 + \sigma_{error}^2$$

Where σ_G^2 is the variance because of genotype, σ_E^2 is the variance because of the environment, σ_{GxE}^2 is the variance because of the genotype by environment interaction and σ_{error}^2 is the variance because of errors. In a controlled F_2 cross, the minimum number of genes controlling a trait (n) can be estimated from the genotypic variance [see below]:

$$n = D^2/8\sigma^2_g$$

where D = the difference between the trait means for the two true-breeding parents of the F_1 plant and σ^2_g is the genotypic variance in the F_2 population (Doerge et al. 1997). While this is a fairly simple calculation that can give us some information about complex inheritance, it cannot give us the locations of the estimated genes nor on how they interact with each other. Indeed, to complicate matters, genes can interact with each other in multiple ways (additive, dominant or a number of epistatic effects) analogous to how different alleles at a single locus can interact with each other.

The multiple genes that control quantitative traits can be difficult to analyze and are nearly impossible to do so using simple Mendelian genetic analysis, even in the simplest cases. To illustrate this problem, let us consider the simplest type of gene interaction, the additive interaction, where each gene adds something to the phenotype. Specifically, let us say that for two genes that affect the same phenotype, A and B, the dominant allele adds a +1 to the phenotype and a recessive allele adds nothing. Let us then self a dihybrid pepper plant, AaBb, and examine the phenotype of the progeny. A Punnett square (Fig. 5-1) gives us all possible combinations of phenotypes. These phenotypes can be sorted into a frequency distribution (Fig. 5-2).

It is still possible to distinguish these different phenotypic classes from each other, but it is not possible to strictly identify the genotype based on

	AB	Ab	aB	ab
AB	4	3	3	2
Ab	3	2	2	1
aB	3	2	2	1
ab	2	1	1	0

Figure 5-1 Punnett square showing phenotypic results of a dihybrid self with a simple quantitative trait showing additive gene interaction (see text).

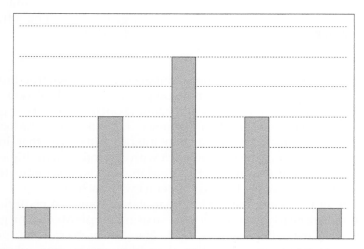

Figure 5-2 Frequency distribution of a dihybrid self where the trait is controlled by two genes with additive gene interaction.

the phenotype, since multiple genotypes can give you the same phenotype. If three genes are involved, the picture becomes even more complex (Fig. 5-3).

Imagine what this would look like if the trait had a heritability of less than 100 percent so that the environment played a role in expression of that phenotype. Different discrete phenotypic classes would then blend together, making it impossible to even identify the specific phenotypic class, much less the underlying genotype (Fig. 5-4). Notice how even with only three genes and a small environmental variance the trait appears essentially continuous (and in this case, a Normal distribution).

The use of genetic markers has enabled investigators to partition the genome in such a way as to identify the loci controlling some of the quantitative traits by using various statistical techniques to search for correlations between these markers and any phenotypes of interest. These statistically identified regions influencing the phenotype of interest are called quantitative trait loci or QTLs. As molecular markers became available and as their numbers increased our abilities to identify QTLs and delimit their locations with higher resolution increased as well. Several characteristics of QTLs controlling traits of interest can be identified through their analysis using molecular maps, including (a) the number of QTLs controlling the trait of interest in the given cross, (b) the locations of those QTLs on the chromosomes and (c) QTL action when compared with other QTLs (Paterson et al. 1991). Increased accuracy of characterization of QTLs is achieved with increased population size and increased numbers of molecular markers.

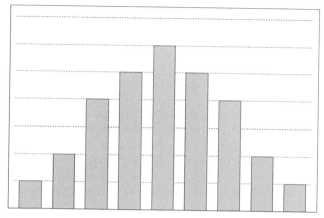

Figure 5-3 Frequency distribution of a trihybrid self where the trait is controlled by three genes with additive gene interaction.

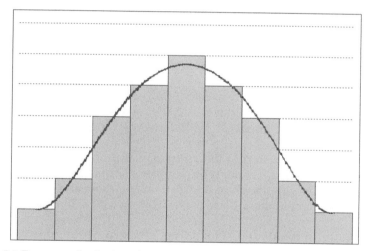

Figure 5-4 Frequency distribution of a trihybrid self-cross where the trait is controlled by three genes with additive gene interaction as well as a noticeable environmental effect.

It is important to realize that, in most cases, QTLs detected are only a statistical estimate of the location of gene(s) controlling the trait of interest. A few genes conferring the phenotype associated with a given QTL have been cloned (see Mendelization of QTL section of Chapter 5).

More information on how to analyze QTLs will be provided later. First, we want to take a snapshot look at the QTL that have been analyzed so far in *Capsicum*. Then we will move on to how these analyses are done.

3. Quantitative Traits Analyzed in Pepper

This section is not meant to be an exhaustive list of all of the QTLs analyzed in pepper, but rather a glimpse at the range of phenotypes that have been analyzed. Based on the effort put into each category, QTL analysis in pepper can be broken down into essentially two categories: (1) disease resistance genes and (2) everything else.

Much of the QTL analysis efforts in pepper have been directed towards disease resistance. QTLs have been identified for resistance to anthracnose (Voorips et al. 2004), cucumber mosaic virus (Caranta et al. 1997b, 2002; Pfleiger et al. 1999; Ben Chaim et al. 2001a), Phytophthora blight (Pflieger et al. 2001; Thabuis et al. 2003; Ogundiwin et al. 2005a; Bonnet et al. 2007; Minamiyama et al. 2007), powdery mildew (Lefebvre et al. 2003) and potyviruses (Caranta et al. 1997a,c; Ogundiwin et al. 2005b). It is becoming clear that some of these resistance QTLs and their candidate genes overlap (Grube et al. 2000; Pflieger et al. 1999, 2001). Does this mean that scientists have identified clusters of individual genes that each confer resistance to different pathogens or isolates of the same pathogen or whether they have found single genes that confer general pathogen resistance? It is not yet clear. It is possible that pepper uses both strategies to combat disease.

Some of the other traits for which QTLs have been mapped in pepper include fruit shape (Ben Chaim et al. 2001b, 2003; Zygier et al. 2005; Jorgensen and Prince 2009), fruit color (Thorup et al. 2000; Huh et al. 2001), fruit capsaicinoid content (Blum et al. 2003; Ben Chaim et al. 2006), number of pedicels per node (Prince et al. 1993), fertility (Wang et al. 2004) and yield (Rao et al. 2003).

Throughout these studies in *Capsicum*, was there a preference for the type of population used of the type of molecular marker, or the analytical software? No! A wide range of populations, markers and software were used. We will discuss these issues next. How are QTLs analyzed?

4. How do you Analyze QTLs using Modern Molecular Techniques?

There are four key issues to consider in setting up a QTL analysis: (a) the phenotype (and its reproducibility), (b) the molecular markers to be used, (c) the populations (and linkage maps) to be analyzed and (d) the analytical software that will identify the QTLs on the basis of linkage between segregant phenotypes and molecular marker genotypes.

4.1 Phenotypic Issues

The essential question about the phenotype is whether or not it shows sufficient variation. Without variation, no analyses are possible. Is variation between two mapping parents necessary for an analysis to proceed, or variation in the segregating population, or both? While having a large phenotypic difference between mapping parents can at least identify a trait of interest that might be amenable to analysis, it is not necessary. It is possible to have similar phenotypes but very different genotypes; this information is hidden when only comparing two individuals at the phenotypic level and would not be revealed unless additional genetic or molecular analyses are performed to see those differences. For instance, consider again our example of genes A and B, both involved in controlling the same phenotype. Using the additive model previously mentioned, the following individuals would be indistinguishable in phenotype:

aaBB and AAbb

$(0 + 0 + 1 + 1)(1 + 1 + 0 + 0)$

Both individuals have a phenotypic score of +2, so they are indistinguishable for this trait. Even the F_1 has the same phenotype (see below). Does this mean that no meaningful genetic analysis of this trait is possible? Of course not. The standard pattern of generating an F_1 individual and the corresponding F_2 would give us a segregating population that can be analyzed (Fig. 5-5). Notice that here each segregant has a discrete value, which does not often happen with quantitative traits because of environmental effects as mentioned previously.

This would be the right time to address the issues of broad-sense heritability according to which the proportion of the total variation seen in a population's phenotype is due to genotypic variance. For QTL analyses

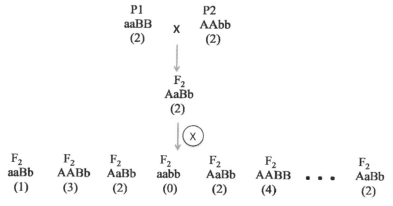

Figure 5-5 Genotypes and phenotypes of the parents, F_1, and F_2 individuals described above in the text. Phenotypic scores are listed in parentheses for the parents, F_1, and representative F_2 individuals. Note that the parents and F_1 have the same phenotype, but you see segregation of the phenotype in the F_2.

to be effective, the heritability of the trait must be large enough so that you can detect genetic effects. If all of the variation seen is due to the environment, there will be no correlations seen between marker genotypes and the phenotype, and therefore no QTL will be identified. How low can heritability go and still afford the chance of detecting those genes that actually do affect the trait? It depends on the resolution of the molecular linkage map, the number of progeny in the mapping cross and whether or not individual QTL have a large effect (i.e., whether you have a "major QTL" rather than just a series of "minor QTLs").

As in most science, reproducibility is the key to confidence in QTL results. Can the same QTL be detected in multiple replicates of the same experiment? Can the same QTL be detected in populations of peppers grown in different locations, at different times and by different investigators? Other types of replication are possible, too, such as replication of infection using multiple isolates of the same species of pathogen. Disease resistance QTL that are detected by challenging the plants with more than one isolate of a pathogen inspire a level of confidence greater than those detected by a single isolate, even though it is becoming clear that the inheritance of disease resistance is complex, including QTLs that confer resistance to multiple pathogens and QTLs that seem isolate specific (Ogundiwin et al. 2005a).

4.2 Molecular Markers

A large variety of molecular markers has been used to identify loci on pepper genetic linkage maps, including restriction fragment length polymorphisms (RFLPs), randomly amplified polymorphic DNA (RAPDs), amplified fragment length polymorphisms (AFLPs), simple-sequence repeats (SSRs)/microsatellites, cleaved amplified polymorphisms (CAPs), sequence-characterized amplified regions (SCARs) and single-nucleotide polymorphisms (SNPs)/single feature polymorphisms (SFPs) (Prince and Ogundiwin 2004; Labate et al. 2007). Each of these markers identifies sequence variation between the mapping parents and thus, in the segregants, in a given QTL mapping cross. Which molecular markers should be used in a QTL analysis? The following factors should be taken into consideration when a choice of molecular marker needs to be made: levels of polymorphism detected, how informative is the marker for each individual segregant (is the marker co-dominant or dominant or does it detect a single locus or distinguishable multiple loci?), ease of use of marker system for a given laboratory group, expense and whether or not the marker has been used in previous mapping experiments (this gives additional information because it provides anchors to linkage groups from the other experiments allowing the possibility of map integration). One of the most important considerations is simply which markers are available and are known to be polymorphic in a given cross. An overview of different types of molecular markers follows (see also Chapter 3):

RFLPs (restriction fragment length polymorphisms). RFLPs, first used in genetic mapping by Botstein et al. (1980), were for many years the workhorse of genetic analysis in pepper. The basic steps in RFLP analysis involve DNA extraction, restriction digestion and Southern hybridization using a cloned DNA probe that will detect one or more genetic loci. RFLPs can usually be scored as co-dominant markers. RFLPs have been largely supplanted by PCR-based markers because of the labor and large amounts of DNA required to detect RFLPs. The rest of the markers to be described are PCR-based markers.

RAPDs (randomly amplified polymorphic DNA). This was the first type of PCR-based marker that was used in pepper genetic linkage analysis. It was developed in very similar form by two groups (Welsh and McClelland 1990; Williams et al. 1990). RAPD analysis involves PCR amplification of genomic DNA using a single arbitrary primer (usually a decamer) at low annealing temperatures, resulting in a number of amplicons (from one to several) being detected on a gel using ethidium bromide. RAPDs are generally scored as dominant markers, but can occasionally be scored as co-dominant loci if allelism between bands from different parents can be established. RAPD markers are rarely used nowadays because of the difficulty of ensuring reproducibility of amplification (Whitekus et al. 1994).

AFLPs (amplified fragment length polymorphisms). Amplified fragment length polymorphism (AFLP) analysis was developed by Vos et al. (1995) and has been used a number of times in pepper QTL analyses (Ben Chaim et al. 2001a,b; Ogundiwin et al. 2005a,b; Sugita et al. 2006). AFLP analysis is a combination of PCR and RFLP analysis. The AFLP method involves (1) cutting genomic DNA with two restriction enzymes, (2) ligation of double-stranded adapters to the ends of the restriction fragments, (3) amplification of a subset of the restriction fragments using two primers complementary to the adapter and restriction site sequences and extended at their 3' ends by "selective" nucleotides and (4) analysis of the amplification products on sequencing gels or in genetic analyzers. AFLP analysis amplifies many loci in each reaction. They are usually scored as dominant loci, because the large number of amplicons (50–100 per reaction) makes it difficult to identify alleles of the same locus. While AFLP analysis is technically challenging, the reproducibility of the amplifications is excellent, and a large number of loci can be analyzed per reaction.

Microsatellites or *SSRs (simple-sequence repeats).* SSRs consist of tandem repeats of less than 6 bp [e.g., (CA)n or (GATA)n] and are synonymous with the term microsatellites (Litt and Luty 1989). These are scattered through most eukaryotic genomes and are often highly polymorphic in repeat copy number, because of unequal crossing-over. They are also usually codominant and multiallelic and therefore serve as excellent genetic markers. They can be detected on gels or in genetic analyzers. SSRs have been used successfully in linkage mapping (Kang et al. 2000; Lee et al. 2004) and QTL analysis in pepper (Minamiyama et al. 2007). Initially, SSR discovery involved library construction and Southern analysis using probes consisting of the desired repeat, followed by sequencing of the selected clones to design specific PCR primers flanking the non-unique SSR sequence. This was very expensive and time-consuming, but the recent availability of EST sequences in pepper from Doil Choi's lab (Lee et al. 2004; Yi et al. 2006) should allow the discovery of a large number of gene-resident SSRs.

SNPs (single-nucleotide polymorphisms). SNPs, naturally occurring single-base differences, are the most common type of variation occurring at the molecular level and the basis for most of the polymorphism detected by the marker systems discussed above. SNP discovery and analysis can be done in many ways, including DNA sequencing, PCR-RFLP (Wicks et al. 2001); allele-specific hybridization (Stoneking et al. 1991); single-stranded conformation polymorphism (SSCP) analysis (Xu et al. 2009) denaturing HPLC (Underhill et al. 1997) and DNA melting analysis (Wittwer 2009). Allen van Deynze and colleagues at UC Davis have just completed a very large tiling microarray experiment based primarily on the EST sequences generated by Doil Choi's lab, and the Davis group has identified many thousands of SFPs (single feature polymorphisms, analogous to SNPs, but perhaps also being a small insertion of deletion) in *Capsicum* and constructed a very-high-resolution linkage map with 4,000+ SFP markers (Hill et al. 2010). Many of these techniques can be automated, meaning SNPs will continue to be an increasingly important molecular marker tool for future genetic mapping and allelic discrimination studies. (See Chapter 3 for a discussion of SNP resources in pepper.)

SCARs (sequence-characterized amplified regions). SCARs (Paran and Michelmore 1993) are sequence-tagged site (STS) with primers designed to amplify a specific DNA sequence. They originated as a way to more reliably and specifically amplify loci of interest that had been identified through RAPD analysis, which was unreliable and amplified multiple loci. In short, the desired RAPD (or other) amplicon was cut from the gel, clones and sequenced. Specific primers were then designed from that sequence with the hope that the same locus would be much more reliably and sensitively amplified. This has worked in large part, and SCAR markers appear on several linkage maps of pepper (Arnedo-Andres et al. 2002; Ogundiwin et al. 2005a; Quirin et al. 2005; Wang et al. 2009; Kumar et al. 2009), but sometimes SCAR markers do not map to the same locus as the originating DNA fragment. Therefore, co-localization should be confirmed and not assumed.

CAPs (cleaved amplified polymorphisms) and dCAPs (derived CAPs). CAPs are PCR amplicons that are subsequently cleaved with restriction enzymes in order to reveal the polymorphisms. Initially, lab scientists would digest monomorphic pairs of amplicons from mapping parents with as many restriction enzymes as they could obtain and hope for polymorphism. Now, it is more common to clone and sequence the parental amplicons and let recognition-site-finding software discover the CAPs. CAPs have been widely used to tag genes in pepper (Yeam et al. 2005; Lee et al. 2008; Lee et al. 2010). Often, sequencing of amplicons reveals SNPs that are not in restriction enzyme recognition sequences. Neff et al. (1998) developed the dCAPs (derived CAPS) marker system, where PCR primers are designed specifically such that amplification of the two different parental amplicons leads to an engineered restriction enzyme recognition sequence in one amplicon and not in the other. This technique is currently being exploited by pepper labs struggling with low intraspecific polymorphism (Lang et al. 2009).

4.3 Populations

Both the type of mapping population and the size of that population can play a large role in the success of QTL mapping. A number of different types of populations can be used for genetic linkage mapping and QTL analysis (see also Chapter 3 of this volume). The key is that the traits of interest, the genes controlling those traits and the molecular markers that will be used for map construction segregate during sexual recombination in the construction of the mapping population. This means that genes assort independently from each other if on different chromosomes and that they can be separated from each other by crossing-over if on the same chromosome. The following types of populations have been used extensively in pepper genetics in the last 15-20 years: F_2, backcross, recombinant inbred lines (RILs) and doubled-haploid (DH) lines. Many molecular genetic linkage maps have been constructed in pepper. Some of those are shown in Table 5-1, along with the type of mapping population and the parents of the populations. Notice that some pepper genotypes have been used multiple times as parents, usually because they have phenotypic traits of interest to the investigators. Some of these populations, the RIL and DH, may be useful for comparative QTL studies for many years, as these populations are highly homozygous and thus immortal, we can reproduce the individual segregant plants indefinitely through selfing, distributing the seeds to a wide number of collaborating labs over long periods of time. This is very helpful for QTL studies, as traits

Table 5-1 A representative sample of types of populations used to construct linkage maps in *Capsicum*. Many of these maps have been used for QTL analysis as well (BC = backcross, RIL = recombinant inbred lines, DH = doubled haploid). Parents are *C. annuum* unless otherwise stated. Notice that some parents and populations have been used for multiple linkage map publications.

Population	Parents	Reference
F2	NuMex R. Naky x *C. chinense* PI 159234	Prince et al. 1993
	Early Calwonder (ECW) x ECS-1123R	Tai et al. 1999
	NuMex R. Naky x *C. chinense* PI 159234	Livingstone et al. 1999
	PM 687 x Yolo Wonder	Djian-Caporalino et al. 2001
	TF68 x *C. chinense* Habanero	Kang et al. 2000
	Yolo Wonder x CM334	Pflieger et al. 2001
	Yolo Wonder x CM334	Lefebvre et al. 2003
	TF68 x *C. chinense* Habanero	Lee et al. 2004
	NuMex R. Naky x *C. chinense* PI1 59234	Paran et al. 2004
	Maor x Perennial	Paran et al. 2004
	Yolo Wonder x CM334	Paran et al. 2004
	CM334 x NuMex Joe E Parker	Ogundiwin et al. 2005a
	TF68 x *C. chinense* Habanero	Yi et al. 2006
	C. frutescens BG2814-6 x NuMex R Naky	Wu et al. 2009
BC	Early Calwonder (ECW) x ECS-1123R	Tai et al. 1999
	100/63 x *C. chinense* PI 152225	Paran et al. 2004
RIL	PI201234 x Psp-11	Ogundiwin et al. 2005a
	CM334 x Yolo Wonder	Barchi et al. 2007
	C. frutescens BG2814-6 x NuMex R Naky	Hill et al. 2010
DH	Perennial x Yolo Wonder	Lefebvre et al. 1996
	Vat x CM334	Lefebvre et al. 1996
	Yolo Wonder x CM334 and others	Lefebvre et al. 1996
	Perennial x Yolo Wonder	Caranta et al. 1997ab
	PM 687 x Yolo Wonder	Djian-Caporalino et al. 2001
	Perennial x Yolo Wonder	Djian-Caporalino et al. 2001
	H3 x Vania	Pflieger et al. 2001
	Yolo Wonder x CM334	Pflieger et al. 2001
	H3 x Vania	Lefebvre et al. 2003
	Perennial x Yolo Wonder	Lefebvre et al. 2003
	H3 x Vania	Paran et al. 2004
	Perennial x Yolo Wonder	Paran et al. 2004
	Manganji x Tongari	Minamiyama et al. 2006

can be analyzed in a larger number of locations, in different seasons and under different environmental conditions.

The choice of parents influences how much polymorphism is seen. It is no surprise that the interspecific crosses show a much higher percentage of polymorphism, and that some of the most recent high-throughput work is being done on interspecific crosses. While maximizing polymorphism for the molecular markers and the traits of interest is important, it can also lead to problems. Crosses between highly genetically divergent individuals can result in decreased fertility or sterility of the F_1 generation preventing the production of F_2 or BC populations. Also, skewed segregation ratios and false linkage assessments can result. Finally, it may be difficult to extrapolate information about a trait of interest from an interspecific cross to what we would normally see in intraspecific breeding programs of elite germplasm.

Efforts have been made to integrate the various linkage maps of pepper with each other as well as with the linkage map of tomato in order to allow for easier transfer of information from lab to lab and crop to crop. This helps in the practical arena of QTL mapping as well as in efforts to understand plant genome evolution. The main efforts in this area have been provided by Livingstone et al. (1999) and Paran et al. (2004), where anchor markers (molecular markers mapped in more than one population) are used in order to determine which parts of linkage maps from one population correspond to which regions in another. Recently, Wu et al. (2009) mapped a set of conserved orthologous sequence (COSII) markers onto an interspecific map of pepper. These markers' positions on the tomato and *Aradibopsis* genome are already known. Hill et al. (2010) have recently produced a gene-chip-SNP map of the pepper genome with over 4,000 markers. Sequences of the markers will provide for significant map integration opportunities in the near future, and this map may very well be the new gold standard for QTL work and in finding markers for integrative mapping in pepper for many years.

The size of a mapping population also has significant effect on the effectiveness of the map generated. In general, the larger the population, the better the map resolution will be, since there will be a greater chance of detecting rare recombination events. Small population size can cause unresolved linkage when a large number of markers are used (Liu 1998).

In most cases, individual plants from a mapping population are selected randomly for analysis. Sometimes it is advantageous to select a smaller set of individuals on the basis of phenotypic or genotypic criteria. Lander and Botstein (1989) describe a QTL detection methodology where measurements of traits are taken over a whole population, but molecular marker genotyping is performed only on those individuals in the extreme tails of the phenotypic distribution. The savings in time and resources of this technique are obvious. Barchi et al. (2007) demonstrated the utility of this technique for "fast mapping" in pepper, by selecting a subset of a larger set of RILs with which to map the genome.

4.4 Details on QTL Analyses

Three methods for detecting and mapping QTLs in a genome are widely used at present: single-marker analysis, interval mapping and composite interval mapping. We will briefly discuss each of these methods. Then we will address the number of molecular markers and the number of segregants needed for QTL analysis, some heritability considerations and a brief look at the software used in order to analyze QTLs in pepper.

4.4.1 Single-Marker Analysis for QTL Mapping

The simplest method for QTL mapping is the utilization of single molecular markers for detecting correlation with phenotypic variation in a population. This method is called simple interval mapping (SIM), and it may involve analysis of variance, linear regression or correlation tests.

For example, a quantitative trait Q/q with a population mean μ is linked with a molecular marker A/a by r recombination units on a linkage group as shown in Fig. 5-6. A linear regression, ANOVA, or correlation analysis between the phenotypic values and the genotypes of each marker on the map for each segregant can identify linkages. A pre-existing linkage map of the chromosomes is not required for this technique to be successful. The single marker approach for QTL mapping has three important weaknesses: (1) the phenotypic effects of QTLs are underestimated, (2) QTL locations are at a low resolution and (3) the number of progeny needed for detection is larger than with more sophisticated techniques (Lander and Botstein 1989). Therefore, an interval approach is much more commonly used for estimating the position of a QTL within two markers, but it requires special software (see Chapter 5, section 4.4.5).

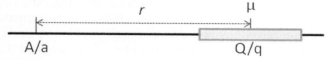

Figure 5-6 Single-marker model for QTL mapping.

4.4.2 Interval Analysis for QTL Mapping

This method is based on maximum likelihood techniques (Lander and Botstein 1989; Jansen 1993), with the assumption that single QTLs are present in the genome between markers. It makes use of existing linkage maps, with an increasing density of markers providing for higher resolution of QTL positions. Interval mapping allows for the estimation of QTL positions and peaks between markers, rather than simply at the marker positions, like single-marker analyses.

Suppose that a QTL is linked with marker A at a recombination frequency of r1 and with B at a recombination frequency of r2 as shown in Fig. 5-7. The recombination frequency among the QTL and the two flanking markers A and B is evaluated using a likelihood function, which is defined as a base 10 logarithm of a likelihood ratio:

$$LOD = \log_{10}(L1/L2)$$

where L1 is the likelihood of the QTL being in the interval between A and B and L2 is the likelihood of the linkage map and L2 is the likelihood of the QTL being in any other interval (or that the QTL appears to be between A and B because of random chance) (Lander and Botstein 1989). If the LOD score for a QTL exceeds a pre-determined threshold over a certain region of the genome, then a QTL is said to exist over that region. An LOD (logarithm of the odds ratio) score ≥ 3 is commonly used as the minimum requirement to support a linkage association of a QTL within an interval, which would mean that the QTL is 1000 times more likely to be in this interval than in other intervals. The actual LOD threshold to be chosen depends on the size of the genome and marker density (Lander and Botstein 1989) and can be determined through permutation analysis (Wang 2007). QTL analysis programs that use interval mapping will produce plots of the LOD scores for a QTL against the positions of molecular markers on a linkage group, as shown in Fig. 5-8. The dotted line and horizontal bar with a triangle at the top were added after the LOD plot was made. The dotted line indicates an LOD significance threshold of 3.0. The horizontal bar indicates that stretch of the chromosome for which a QTL is detected above the threshold and the point on the triangle corresponds to the peak of the QTL. Whole genome QTL scans for multiple traits can also be visualized, as shown in Fig. 5-9, where the entire pepper genome has been scanned for QTLs controlling 12 horticultural traits of interest.

Interval mapping can also determine whether or not individual QTL are acting in additive or dominant fashions. Epistatic interactions often need to be detected by using pairwise regressions with SAS (statistical analysis system) or a similar statistical package. While interval mapping is clearly advantageous over single-marker mapping, it may misinterpret linked QTLs as being a single strong QTL (Zeng 1994). Composite interval mapping was developed to address this problem.

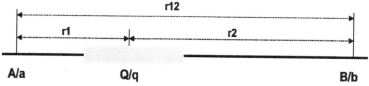

Figure 5-7 Multiple marker model for interval mapping of QTLs.

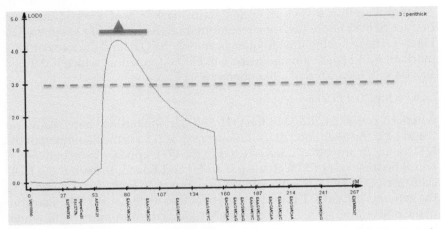

Figure 5-8 LOD score tracing from QTL Cartographer of a QTL on one linkage group in the pepper genome (Jorgensen C and Prince JP, unpublished data).

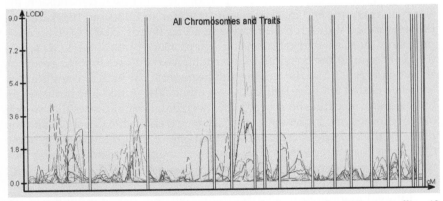

Figure 5-9 Whole-genome scan of pepper by QTL Cartographer for QTLs controlling 12 horticulturally important traits in pepper (Jorgensen C and Prince JP, unpublished results). Notice the LOD score significance threshold below 3.0.

Color image of this figure appears in the color plate section at the end of the book.

4.4.3 Composite Interval Analysis for QTL Mapping

Composite interval mapping is a technique developed simultaneously by two different groups (Jansen and Stam 1994; Zeng 1994) that combines standard interval analysis with multiple linear regression methods using appropriate unlinked markers as cofactors in the analysis. This allows a higher resolution of QTL mapping and a reduction in the genetic "background noise" to enable the detection of true QTLs that have smaller effects.

4.4.4 Number of Markers Required for QTL Mapping

How many markers does a genetic linkage map need for successful QTL mapping? Several methods have been used to estimate this quantity. In one early strategy from human genome work, Bishop et al. (1983) developed a relationship between the number of markers (n), genome size (C), total map units (L) and the fraction of all loci within a certain number of map units of any given marker (m):

$$p = 1 - 2C\,[(1-x)^{n+1} - (1-x)^{n+1}]/(n+1) - (1-2xC)\,(1-2x)^n$$

where x = m/L, n is the number of random markers.

The number of markers required for QTL mapping varies greatly with genome size as well as the amount of recombination. Usually, 100 to 200 markers are enough to construct a genome 'framework' for detection of QTLs in many plant species (Mohan et al. 1997). This translates into molecular markers with a genetic spacing of 10–20 cM on average having enough power of detecting a QTL (Darvasi et al. 1993). Naturally, more markers will give better results. Often, the number of markers is determined more by personnel and resource limitations than by theoretical considerations, but the interpretation of the data must certainly take this into account.

4.4.5 QTL Mapping Software

Software for QTL analyses varies widely. All of them need, in common, is sets of molecular marker genotypic data, all in the same order of segregant for each locus, as well as a corresponding set of phenotypic data, in the same order as the molecular data. If you have data that are out of order, your analyses will not work. It is critical to obtain a copy of the manual (written or electronic) for the software that will be used for QTL analyses. In particular, it is easy to format the input data incorrectly, and when you move from platform to platform, it is often necessary to completely reformat the data. This can be frustrating.

The programs that have primarily been used in QTL analysis projects in pepper include Mapmaker/QTL (Lander and Botstein 1989), QTL Cartographer (Basten et al. 2002), MapQTL 4.0 (Van Ooijen 2002), QGene (Nelson 1997), R/QTL (Broman et al. 2003) and SAS (SAS Institute 1989). Some programs work better than others for certain applications, and it is necessary for each research team to choose the one that best fits their needs. Table 5-2 gives the current web address for obtaining each program as of late 2010.

Table 5-2 Websites to obtain QTL analysis software.

QTL Program	Website
Mapmaker/QTL	http://www.broadinstitute.org/scientific-community/software*
QTL Cartographer	http://statgen.ncsu.edu/qtlcart/
MapQTL 4.0	http://www.kyazma.nl/index.php/mc.MapQTL
QGene	http://coding.plantpath.ksu.edu/qgene/
R/QTL	http://www.rqtl.org/
SAS	http://www.sas.com/software/sas9/

*Type "Mapmaker" into the dialog box. "Mapmaker3" will pop up. Download this. It includes Mapmaker/EXP Version 3.0 and Mapmaker/QTL Version 1.1

5. Mendelization of QTLs and Marker-Assisted Selection (MAS)

QTLs have been successfully treated as individual genes in breeding programs, using the principles of marker-assisted selection (MAS), where multiple QTLs controlling a trait are separated from each other by following the segregation of molecular markers closely linked to each QTL through a breeding program, and selecting for the combination of markers, and thus QTLs, that you are seeking (see section 4.4 of this chapter). The technique of marker-assisted selection of QTLs (MAS-QTL) is specifically described by Collard et al. (2005). As early as in 1996, this technique was successful in separating individual tomato QTLs from each other (Tanksley and Nelson 1996). In a series of papers published by Thabuis and colleagues in France (Thabuis et al. 2001, 2004a,b), showed successful marker-assisted selection programs for several *Phytophthora capsici* resistance QTLs in pepper, and Quirin et al. (2005) developed a SCAR marker to tag the *P. capsici* resistance QTL *Phyto5.2*, potentially Mendelizing that QTL for future breeding efforts. Efforts are currently underway led by the laboratory of Allen van Deynze at UC Davis, to develop ultra-high density QTL maps of *P. capsici* resistance and other traits (Hill et al. 2010; Van Deynze, personal communication). Several of the large seed companies have invested heavily for a number of years in high-density pepper maps and QTL analyses, as well, but this information is proprietary. These ultra-high density maps will enable the ready partitioning of individual gene action for a number of complex traits.

Acknowledgements

We would like to thank the members of the Prince Lab for their hard work and the NIH, USDA, California State University Agricultural Research Initiative (CSU-ARI), the California Pepper Commission, the California State University Program for Education and Research in Biotechnology

(CSUPERB) and the CSU Fresno College of Science and Mathematics for funding our research.

References

Andrews J (1984) Peppers: The Domesticated Capsicums. University of Texas Press, Austin, USA.

Arnedo-Andres MS, Hormaza JL, Luis-Arteaga M, Hormaza J (2002) Development of RAPD and SCAR markers linked to the Pvr4 locus for resistance to PVY in pepper (*Capsicum annuum* L.). Theor Appl Genet 105: 1067–1074.

Barchi L, Lefebvre V, Sage-Palloix AM, Lanteri S, Palloix A (2007) QTL analysis of plant development and fruit traits in pepper and performance of selective phenotyping. Theor Appl Genet 118: 1157–1171.

Basten CJ, Weir BS, Zeng ZB (2002) QTL CARTOGRAPHER, Version 1.15. Raleigh, NC, USA, Department of Statistics.

Ben Chaim A, Borovsky Y, De Jong W, Paran I (2003) Linkage of the A locus for the presence of anthocyanin and fs10.1, a major fruit-shape QTL in pepper. Theor Appl Genet 106: 889–894.

Ben Chaim A, Borovsky Y, Falise M, Mazourek M, Kang BC, Paran I, Jahn M (2006) QTL analysis for capsaicinoid content in *Capsicum*. Theor Appl Genet 113: 1481–1490.

Ben Chaim A, Grube RC, Lapidot M, Jahn M, Paran I (2001a) Identification of quantitative trait loci associated with resistance to cucumber mosaic virus in *Capsicum annuum*. Theor Appl Genet 102: 1213–1220.

Ben Chaim A, Paran I, Grube RC, Jahn M, Wijk van R, Peleman J (2001b) QTL mapping of fruit-related traits in pepper (*Capsicum annuum*). Theor Appl Genet 102: 1016–1028.

Bishop DT, Cannings C, Skolnick M, Williamson JA (1983) The number of polymorphic DNA clones required to map the human genome. In: Weir BS [ed] Statistical Analysis of DNA Sequence Data, Marcel Dekker, New York, USA pp 181–200.

Blum E, Mazourek M, O'Connell M, Curry J, Thorup T, Liu K, Jahn M, Paran I (2003) Molecular mapping of capsaicinoid biosynthesis genes and quantitative trait loci analysis for capsaicinoid content in *Capsicum*. Theor Appl Genet 108: 79–86.

Bonnet J, Danan S, Boudet C, Barchi L, Sage-Palloix AM, Caromel B, Palloix A, Lefebvre V (2007) Are the polygenic architectures of resistance to *Phytophthora capsici* and *Phytophthora parasitica* independent in pepper? Theor Appl Genet 115: 253–264.

Botstein D, White RL, Skolnick MH, Davis RW (1980) Construction of a genetic linkage map in man using restriction fragment length polymorphisms. Am J Hum Genet 32: 314–331.

Broman KW, Wu H, Sen Ś, Churchill GA (2003) R/qtl: QTL mapping in experimental crosses. Bioinformatics 19: 889–890.

Caranta C, Lefebvre V, Palloix A (1997a) Polygenic Resistance of Pepper to Potyviruses Consists of a Combination of Isolate-Specific and Broad-Spectrum Quantitative Trait Loci. Mol Plant-Microbe Interact 10: 872–878.

Caranta C, Palloix A, Lefebvre V, Daubeze AM (1997b) QTLs for a component of partial resistance to cucumber mosaic virus in pepper: restriction of virus installation in host-cells. Theor Appl Genet 94: 431–438.

Caranta C, Pflieger S, Lefebvre V, Daubeze AM, Thabuis A, Palloix A (2002) QTLs involved in the restriction of cucumber mosaic virus (CMV) long-distance movement in pepper. Theor Appl Gene 104: 586–591.

Collard BCY, Jahufer MZZ, Brouwer JB, Pang ECK (2005) An introduction to markers, quantitative trait loci (QTL) mapping and marker-assisted selection for crop improvement: the basic concepts. Euphytica 142: 169–196.

Darvasi A, Weinreb A, Minke V, Weller JI, Soller M (1993) Detecting marker-QTL linkage and estimating QTL gene effect and map location using a saturated genetic map. Genetics 134: 943–951.

Djian-Caporalino C, Pijarowski L, Fazari A, Samson M, Gavean L, O'Byrne C, Lefebvre V, Caranta C, Palloix A, Abad P (2001) High-resolution genetic mapping of the pepper (*Capsicum annuum L.*)resistance loci Me3 and Me4 conferring heat-stable resistance to root-knot nematodes (Meloidogyne spp.). Theor Appl Genet 103: 592–600.

Doerge RW, Zeng Z-B, Weir B (1997) Statistical issues in the search for genes affecting quantitative traits in experimental populations. Statist Sci 12: 195–219.

Grube RC, Blauth JR, Arnedo-Andres M, Caranta C, Jahn MK (2000) Identification and comparative mapping of a dominant potyvirus resistance gene cluster in *Capsicum*. Theor Appl Genet 101: 852–859.

Hill T, Ashrafi H, Chin-Wo SR, Kozik A, Van Deynze A (2010) Ultra high density EST-based maps reveal genome differences between *C. frutescens* and *C. annuum*. Proceedings, Plant and Animal Genome XVIII. San Diego, CA, Poster abstract.

Huh JH, Kang BC, Nahm SH, Kim S, Ha KS, Lee, Kim BD (2001) A candidate gene approach identified phytoene synthase as the locus for mature fruit color in red pepper *(Capsicum spp.)*. Theor Appl Genet 102: 524–530.

Jansen RC (1993) Interval mapping of multiple quantitative trait loci. Genetics 135: 205–211.

Jansen RC, Stam P (1994) High resolution of quantitative traits into multiple loci via interval mapping. Genetics 136: 1447–1455.

Jorgensen C, Prince JP (2009) Abstract. QTL analysis and marker addition in a recombinant inbred line of pepper. Proceedings, Plant & Animal Genome XVII.

Kang BC, Nahm SH, Huh JH, Yoo HS, Kim BD (2000) Construction of a molecular linkage map in hot pepper. Acta Horticulturae 521: 165–172.

Kumar R, Singh M, Yaday DS, Raj M, Raj A, Kumar S, Dwivedi N, Kumar S (2009) Validation of SCAR markers, diversity analysis of male sterile (S-) cytoplasm and isolation of an alloplasmic S-cytoplasm in *Capsicum*. Scientia Horticulturae 120: 167–172.

Labate JA, Grandillo S, Fulton T, Muños S, Caicedo AL, Peralta I, Ji Y, Chetelat RT, Scott JW, Gonzalo MJ, Francis D, Yang W, van der Knaap E, Baldo AM, Smith-White B, Mueller LA, Prince JP, Blanchard NE, Storey DB, Stevens MR, Robbins MD, Wang JF, Liedl BE, O'Connell MA, Stommel JR, Aoki K, Iijima Y, Slade AJ, Hurst SR, Loeffler D, Steine MN, Vafeados D, McGuire C, Freeman C, Amen A, Goodstal J, Facciotti D, van Eck J, Causse M (2007) Tomato. In: Kole C [ed] Genome Mapping and Molecular Breeding in Plants. Springer, Berlin, Heidelberg, New York.

Lander ES, Botstein D (1989) Mapping Mendelian factors underlying quantitative traits using RFLP linkage maps. Genetics 121: 185–199.

Lang Y, Tanaka Y, Watanabe T, Miwa T, Yazawa S, Sugiyama R, Kisaka H, Morita A, Nomura K (2009) Functional loss of pAMT results in biosynthesis of capsinoids, capsaicinoid analogs, in *Capsicum annuum* cv. CH-19 Sweet. Plant J 59: 953–961.

Lee J, Kim SH, Park HG, Lee WP, Yoon JB, Han JH (2010) Three AFLP markers tightly linked to the genic male sterility ms gene in chili pepper (*Capsicum annuum L.*) and conversion to a CAPS marker. Euphytica 173: 55–61.

Lee JM, Nahm SH, Kim YM, Kim BD (2004) Characterization and molecular genetic mapping of microsatellite loci in pepper. Theor Appl Genet 108: 6719–627.

Lee S, Pai HS, Hur CG, Choi D, Kim SY, Chung E, Joung YH (2004) EST and microarray analyses of pathogen-responsive genes in hot pepper (*Capsicum annuum L.*) non host-resistance against soybean pustule pathogen (*Xanthomonas axonopodis* pv. *glycines*). Funct Integr Genomics 4: 196–205.

Lee J, Park HG, Yoon JB (2008) A CAPS marker associated with the partial restoration of cytoplasmic male sterility in chili pepper (*Capsicum annuum L.*). Mol Breed 21: 95–104.

Lefebvre V, Daubeze AM, van der Voort JR, Peleman J, Bardin M, Palloix A (2003) QTLs for resistance to powdery mildew in pepper under natural and artificial infections. Theor Appl Genet 107: 661–666.

Litt M, Luty J (1989) A hypervariable microsatellite revealed by *in vitro* amplification of dinucleotide repeat within the cardiac muscle actin gene. Am J Hum Genet 44: 397–401.

Liu BH (1998) Statistical Genomics: linkage, mapping, and QTL analysis. CRC Press, Florida, USA.

Livingstone KD, Lackney VK, Blauth JR, van Wijk R, Jahn MK (1999) Genome mapping in *Capsicum* and the evolution of genome structure in the Solanaceae. Genetics 152: 183–202.

Mohan M, Nair S, Bhagwat A, Krishna TG, Yano M, Bhatia CR, Sasaki T (1997) Genome mapping, molecular markers and marker-assisted selection in crop plants. Mol Breed 3: 87–103.

Neff MM, Neff JD, Chory J, Pepper AE (1998) dCAPS, a simple technique for the genetic analysis of single nucleotide polymorphisms: experimental applications in *Arabidopsis thaliana* genetics. Plant J 14: 387–392.

Nelson JC (1997) QGENE: Software for marker-based genomic analysis and breeding. Mol Breed 3: 239–245.

Ogundiwin EA, Berke T, Massoudi M, Black LL, Huestis G, Choi D, Lee S, Prince P (2005a) Construction of 2 intraspecific linkage maps and identification of resistance QTLs for *Phytophthora capsici* root-rot and foliar-blight diseases of pepper (*Capsicum annuum* L.). Genome 48: 698–711.

Ogundiwin EA, Green S, Berke TF, Hanson P, Prince JP (2005b) Abstract. Detection of QTLs controlling resistance to potato virus Y (PVY) and chili veinal mottle virus (ChiVMV) in pepper (*Capsicum annuum* L.). Proceedings, Plant & Animal Genome XIII.

Paran I, Michelmore RW (1993) Development of reliable PCR-based markers linked to downy mildew resistance genes in lettuce. Theor Appl Genet 85: 985–993.

Paran I, van der Voort JR, Lefebvre V, Jahn M, Landry L, Van Schriek M, Tanyolac B, Caranta C, Ben Chaim A, Livingstone K, Palloix A, Pelemen J (2004) An integrated genetic linkage map of pepper. Mol Breed 13: 251–261.

Paterson AH, Tanksley SD, Sorrells ME (1991) DNA markers in plant improvement. Advances in Agronomy 46: 39–90.

Pflieger S, Lefebvre V, Caranta C, Blattes A, Goffinet B, Palloix A (1999) Disease resistance gene analogs as candidates for QTLs involved in pepper-pathogen interactions. Genome 42: 1100–1110.

Pflieger S, Palloix A, Caranta C, Blattes A, Lefebvre V (2001) Defense response genes co-localize with quantitative disease resistance loci. Theor Appl Genet 103: 920–929.

Prince JP, Pochard E, Tanksley SD (1993) Construction of a molecular linkage map of pepper and a comparison of synteny with tomato. Genome 36: 404–417.

Prince JP, Ogundiwin EA (2004) Molecular mapping and marker-assisted selection of quantitative trait loci (QTL) in plants. In: Klee H, Christou P [ed] Handbook of Plant Biotechnology, John Wiley & Sons, Chichester, UK.

Quirin EA, Ogundiwin EA, Prince JP, Mazourek M, Briggs MO, Chlanda TS, Kang BC, Kim KT, Falise M, Jahn MM (2005) Development of sequence characterized amplified region (SCAR) primers for the detection of *Phyto.5.2*, a major QTL for resistance to *Phytophthora capsici* Leon. in pepper. Theor Appl Genet 110: 605–612.

Rao GU, Ben Chaim A, Borovsky Y, Paran I (2003) Mapping of yield-related QTLs in pepper in an interspecific cross of *Capsicum annuum* and *C. frutescens*. Theor Appl Genet 106: 1457–1466.

SAS Institute, Inc (1989) *SAS/STAT user's guide, version 6, 4th edition*. SAS Institute, Inc., Cary, North Carolina, USA.

Stoneking M, Hedgecock D, Miguchi RG, Vigilant L, Erlich HA (1991) Population variation of human mtDNA control region sequences detected by enzymatic amplification and sequence-specific oligonucleotide probes. Am J Hum Genet 48: 370–382.

Sugita T, Yamaguchi K, Kinoshita T, Yuji K, Sugimura Y, Yagata R, Kawasaki S, Todoroki A (2006) QTL analysis for resistance to Phytophthora blight (*Phytophthora capsici* Leon.) using an intraspecific doubled-haploid population of *Capsicum annuum*. Breed Sci 56:137–145.

Tai TH, Dahlbeck D, Stall RE, Peleman J, Staskawicz BJ (1999) High-resolution genetic physical mapping of the region containing the *Bs2* resistance gene of pepper. Theor Appl Genet 99: 1201–1206.

Tanksley SD, Nelson JC (1996) Advanced backcross QTL analysis: a method for the simultaneous discovery and transfer of valuable QTLs from unadapted germplasm into elite breeding lines. Theor Appl Genet 92: 191–203.

Thabuis A, Palloix A, Pflieger S, Daubeze AM, Caranta C, Lefebvre V (2003) Comparative mapping of Phytophthora resistance loci in pepper germplasm: evidence for conserved resistance loci across Solanceae and for a large genetic diversity. Theor Appl Genet 106: 1473–1485.

Thabuis A, Lefebvre V, Bernard G, Daubeze AM, Phaly T, Pochard E, Palloix A (2004a) Phenotypic and molecular characterization of a recurrent selection program for a polygenic resistance to *Phytophthora capsici* in pepper. Theor Appl Genet 109: 342–351.

Thabuis A, Lefebvre V, Daubeze AM, Signoret P, Phaly T, Nemouchi G, Blattes A, Palloix A (2001) Introgression of a partial resistance to *Phytophthora capsici* Leon. into a pepper elite line by marker assisted backcrosses. Acta Horticulturae 546: 645–650.

Thabuis A, Palloix A, Servin B, Daubeze AM, Signoret, Hospital F, Lefebvre V (2004b) Marker-assisted introgression of 4 *Phytophthora capsici* resistance QTL alleles into a bell pepper line: validation of additive and epistatic effects. Mol Breed 14: 9–20.

Thorup TA, Tanyolac B, Livingstone KD, Popovsky S, Paran I, Jahn M (2000) Candidate gene analysis of organ pigmentation loci in the Solanaceae. Proc Natl Acad Sci USA 97: 11192–11197.

Underhill PA, Jim L, Lin HA, Mehdi SQ, Jenkins T, Vollrath D, Davis RW, Cavalli-Sforza LL, Oefner PJ (1997) Detection of numerous Y chromsome biallelic polymorphisms by denaturing high-performance liquid chromoatography. Genome Res 10: 996–1005.

Van Ooijen W (2002) MapQTL® 4.0. *Software for the calculation of QTL position on genetic maps.* Plant Research International. Wagenigen, The Netherlands.

Vos P, Hogers R, Bleeker M, Reijans M, Van de Lee T, Hornes M, Frijters A, Pot J, Peleman J, Kuiper M, Zabeau M (1995) AFLP: a new technique for DNA fingerprinting. Nucleic Acids Res 23: 4407–4414.

Wang S, Basten CJ, Zeng ZB (2007) *Windows QTL Cartographer 2.5.* Department of Statistics, North Carolina State University, Raleigh, North Carolina, USA.

Wang LH, Zhang BX, Lefebvre V, Huang SW, Daubeze AM, Palloix A (2004) QTL analysis of fertility restoration in cytoplasmic male sterile pepper. Theor Appl Genet 109: 1058–1063.

Wang LH, Peng DL, Yun XF, Zhang BX, Zhang ZH, Gu XH, Hua MY, Mao SL (2009) A SCAR marker linked to the N gene for resistance to root knot nematodes *Meloidogyne* spp.) in pepper (*Capsicum annuum* L.). Scientia Horticulturae 2: 318–322.

Welsh J, McClelland M (1990) Fingerprinting genomes using PCR with arbitrary primers. Nucleic Acids Res 18: 7213–7218.

Whitekus R, Doebley J, Wendel JF (1994) Nuclear DNA markers in systematics and evolution. In: Phillips, RL, Vasil IK [ed] DNA-based markers in plants. Kluwer Academic Publishers, Norwell, MA, USA, pp 116–114.

Wicks SR, Yeh RT, Gish WR, Waterson RH, Plasterk RH (2001) Rapid gene mapping in *Caenorhabditis elegans* using a high density polymorphism map. Nat Genet 28: 160–164.

Williams JGK, Kubelik AR, Livak KJ, Rafalski JA, Tingey SV (1990) DNA polymorphisms amplified by arbitrary primers are useful as genetic markers. Nucleic Acids Res 18: 6531–6535.

Wittwer CT (2009) High-resolution DNA melting analysis: advancements and limitations. Hum Mutat 30: 857–859.

Wu F, Eannetta N, Xu Y, Durrett R, Mazourek M, Jahn M, Tanksley SD (2009) A COSII genetic map of the pepper genome provides a detailed picture of synteny with tomato and new insights into recent chromosome evolution in the genus *Capsicum*. Theor Appl Genet 118: 1279–1293.

Xu X, Kawasaki S, Fujimura T, Babu R (2009) A high-throughput, low-cost gel-based SNP assay for positional cloning and marker-assisted breeding of useful genes in cereals. Plant Breeding 128: 325–331.

Yeam I, Kang BC, Lindeman W, Faber N, Frantz JD, Jahn MM (2005) Allele-specific CAPS markers based on point mutations in resistance alleles at the *pvr-1* locus encoding eIF4E in *Capsicum*. Theor Appl Genet 112: 178–186.

Yi G, Lee JM, Lee S, Choi D, Kim BD (2006) Exploitation of pepper EST-SSRs and an SSR-based linkage map. Theor Appl Genet 114: 113–130.

Zeng ZB (1994) Precision mapping of quantitative trait loci. Genetics 136: 1457–1468.

Zygier S, Ben Chaim A, Efrati A, Kaluzky G, Borovsky Y, Paran I (2005) QTL mapping for fruit size and shape in chromosomes 2 and 4 in pepper and a comparison of the pepper QTL map with that of tomato. Theor Appl Genet 111: 437–445.

6

The Structure of Pepper Genome

Minkyu Park and *Doil Choi**

ABSTRACT

Pepper is an important crop because of its unique trait: pungency. Pepper genome sizes range from 3090 Mbp to 5643 Mbp. Initial genomic studies of pepper were carried out by sequencing expressed sequence tags, constructing bacterial artificial chromosome (BAC) libraries and by comparing genetic maps. These tools enabled the development of molecular markers. Karyotyping of pepper chromosomes was performed by fluorescence *in situ* hybridization analysis using rDNA probes. The structure of the pepper genome was analyzed by comparing BAC sequences among Solanaceous species. This revealed the important role of the long-terminal repeat retrotransposon in pepper genome expansion. Recent next-generation sequencing (NGS) techniques have opened up new opportunities to study the pepper genome. The low cost of NGS has enabled the pepper genome project to be launched, thus providing impetus for further study of the pepper genome.

Keywords: Genome, Karyotype, Whole genome sequence (WGS), Expressed sequence tag (EST), Bacterial artificial chromosome (BAC), Long-terminal repeat (LTR)

1. Introduction

Since the completion of the first whole genome sequence (WGS) for *Arabidopsis thaliana*, the number of WGS projects for plant species has radically increased. Whole genome sequences have already been completed for major crop plants such as rice, maize, soybean and potato, and for horticultural plants such as cucumber, grape, apple and papaya.

Department of Plant Science, Seoul National University, Seoul 151-921, Republic of Korea.
*Corresponding author

The genome project for tomato, a model Solanaceous species, has also progressed, and sequencing has recently been completed. A WGS project was recently started for pepper, another Solanaceae species.

Pepper is an important commercial crop because of its unique pungency. Pepper is also widely used as a research material. Until recently, most genomic studies on pepper were carried out using genetic maps. The chromosome numbering of pepper was carried out by comparing the genetic maps of pepper and tomato. Comparative analysis of the genetic maps of pepper and tomato revealed recombinations, inversions and translocations in the pepper genome. The construction of bacterial artificial chromosome (BAC) libraries enabled comparative analyses of pepper and other Solanaceous species at a finer level. The pepper whole genome sequencing project was recently launched and is making progress. This chapter summarizes the prior research on pepper genome and describes the pepper genome sequencing project that is currently in progress.

2. The Estimated Size of the Pepper Genome

The size of the pepper genome was estimated using ethidium bromide flow cytometry. Using this method, Belletti et al. (1998) and Moscone et al. (2003) estimated the genome sizes of 12 *Capsicum* species (Table 6-1). The estimated sizes of the pepper haploid chromosomes ranged from 3090 Mbp to 5643 Mbp. The genome size of the most popular commercial species, *C. annuum*, was the smallest (3090 Mbp) among those measured. The genomes of the cultivated *Capsicum* species, including *annuum, frutescens, chinense,* and *baccatum,* were smaller than those of the wild races.

Table 6-1 The genome sizes of different pepper species.

Genus	Species	1C (Mbp)	Original Reference
Capsicum	*annuum*	3090	Moscone et al. 2003
Capsicum	*frutescens*	3325	Moscone et al. 2003
Capsicum	*chinense*	3345	Moscone et al. 2003
Capsicum	*baccatum*	3628	Moscone et al. 2003
Capsicum	*chacoense*	3746	Belletti 1998
Capsicum	*tovarii*	3878	Belletti 1998
Capsicum	*eximium*	3971	Moscone et al. 2003
Capsicum	*pubescens*	4372	Moscone et al. 2003
Capsicum	*cardenasii*	4386	Belletti 1998
Capsicum	*campylopodium*	4430	Moscone et al. 2003
Capsicum	*praetermissum*	4465	Belletti, 1998
Capsicum	*parvifolium*	5643	Moscone et al. 2003

Among the Solanaceous species, the number of chromosomes is highly conserved at 12. The pepper genome consists of 12 chromosomes and has not undergone whole genome duplication or polyploidization since its speciation within the Solanaceae (Wu et al. 2009). Although the speciation between pepper and tomato occurred only 19.2 million years ago (Wang et al. 2008), the pepper genome is extraordinarily large compared to other Solanaceous species. The average genome size of the other well known Solanaceous species, such as tomato, potato and tobacco, is 1000 Mb, whereas the pepper genome is approximately three times larger.

3. The Available Resources of the Pepper Genome

As an initial effort to study the pepper genome, a BAC library was constructed using *C. annuum*, CM334. This wild race of *C. annuum* has several useful traits, such as resistance to multiple diseases, pungency, trichomes and carotenoid biosynthesis. The estimated genome size of CM334 is 3090 Mb, which is smaller than that of the other *Capsicum* species. The BAC library was constructed using the *Hind*III enzyme, which produced an average insert size of 80 Kb and a total of 78,336 clones. Subsequently, a 12X deep coverage BAC library was constructed for the pepper genome by using the same enzyme with an average insert size of 130 Kb and a total of 235,000 clones (Yoo et al. 2003). Using this library, the BAC clones containing several important genes such as *Capsanthin-Capsorubin Synthase* and *Phytoene Synthase* were screened and useful markers linked to the *C* and *L* loci were developed.

The next effort to study the pepper genome was the sequencing of expressed sequence tags (ESTs). The cDNAs of *C. annuum*, Bukang were sequenced by the Sanger sequencing method, and a total of 122,582 raw EST sequences were generated (Table 6-2) (Kim et al. 2008a). From these sequences, a total of 22,011 unigenes were identified. A database of these sequences was constructed and is publicly available online (http:// genepool.kribb.re.kr). A total of 21 tissues were used for generating the EST sequences (Table 6-3). These tissues include callus and pathogen-infected leaf, in addition to the normal organs such as fruit, flower, leaf and stem.

Table 6-2 Pepper EST sequence statistics.

Data Type	Count
Raw EST	122,585
Pre-processed EST	116,412
Consensus	11,225
Singleton	11,585
Low complexity trimmed singleton	10,786

Tissue-specific expression of ESTs was also analyzed. The results of this analysis are included in the online database. In the functional classification by the Munich Information Center for Protein Sequences (MIPS), the most abundant genes among those classified were related to metabolism (Fig. 6-1). Genes related to transport and subcellular localization were the next most abundant.

Table 6-3 Source tissues and libraries for tissue-specific and total selective EST sequences.

Tissue name	Library description	No. of specific	No. of selective
Pathogen infected leaf		188	241
	KS01: Bacterial_infected_leaf		
	KS13: Pytophthora_Capsici_infected_plant		
	KS26: rbcS-silenced_leaves		
Flower		108	116
	KS07: Flower_Bud		
	KS19: Flower_Bud_size3-8mm		
	KS20: Open_Flower		
Anther		8	11
	KS08: Anther		
Fruit		283	290
	KS09: Young_Fruit		
	KS14: Fruit_immature_1.5-7.5cm+seeds		
	KS15: Fruit_mature_green_pericarp-seeds		
	KS16: Fruit_breaker_pericarp-seeds		
	KS17: Fruit_mature_red_ripe_pericarp-		
Root		86	100
	KS10: Hair_Root		
	KS11: Early_Root		
Placenta		45	62
	KS12: Placenta		
Seed		306	326
	KS18: Seeds_of_immature_and mature_green		
Bark		130	141
	KS21: Bark		
Peduncle		238	259
	KS22: Peduncle		
Callus		87	106
	KS23: Callus_regeneration		
Seedling		231	251
	KS24: Germination_stage_4_and_6_day		
	KS25: Germination_stage_18_day		

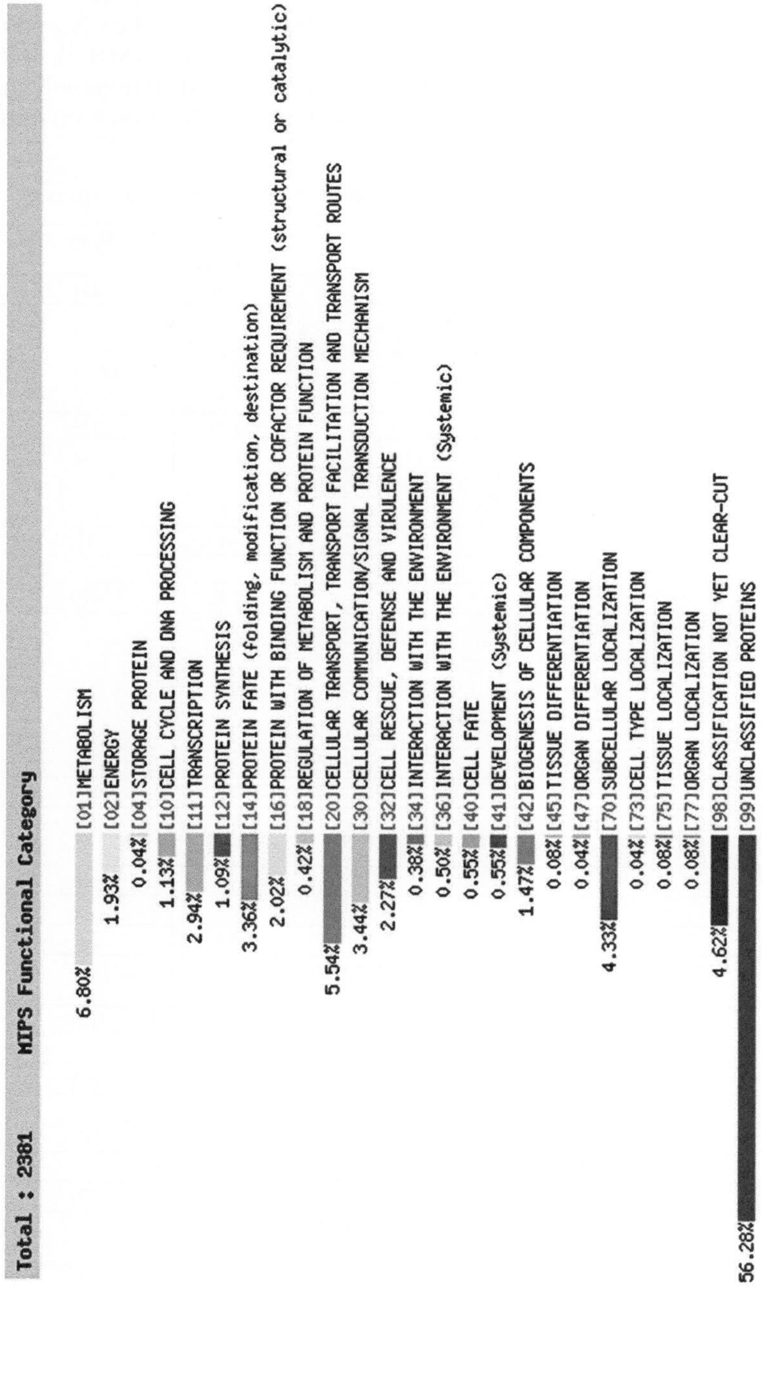

Figure 6-1 MIPS functional categories of the pepper ESTs.

Color image of this figure appears in the color plate section at the end of the book.

4. The Use of BAC Clones in Marker Development

Several studies used the pepper BAC clones to develop markers that are closely linked to useful phenotypes. The long interval between BAC end sequences (approximately 120 Kb) allowed these sequences to be used to find markers that are very close to the target phenotype. Three phenotypic makers in *Capsicum* were reportedly developed using pepper BAC clones.

Kim et al. (2008c) used the BAC clones to develop markers linked to quantitative trait loci (QTLs) controlling resistance to *Phytopthora capsisi*. To find these markers, a molecular linkage map was constructed using 100 F_2 progenies of a cross between *C. annuum*, CM334 (resistant) and *C. annuum*, Chilsungcho (susceptible). The map, which consisted of 209 molecular markers, was used to detect four QTLs explaining 66.3 percent of the total phenotypic variation for root rot resistance and three QTLs explaining 44.9 percent of the total phenotypic variation for damping-off resistance. BAC clones were screened for two of these QTLs and were used to develop single-nucleotide amplified polymorphism markers and cleaved amplified polymorphic sequence markers, which can be used to screen for the *Phytopthora capsisi* resistance phenotype.

Tomita et al. (2009) used BAC clones to develop markers linked to the tobamovirus resistance gene L^3 in *Capsicum chinense*. A total of 2,016 F_2 progenies produced by intra-species crosses and 3,391 F_2 progenies produced by inter-species crosses were used to develop the markers linked to L^3. Analysis of BAC clones with amplified fragment length polymorphism markers linked to L^3 revealed the presence of the homologous gene *I2*: the tomato disease resistance gene. The L^3 gene was mapped to a position between a marker for the *I2* homolog and a BAC-end marker. In this study, the location of L^3 was estimated to be between two BAC contigs separated by an interval of approximately 30 Kb.

Kim et al. (2008b) used the BAC clones to analyze the pepper trichome locus 1 (*Ptl1*) in *C. annuum*, CM334. A total of 653 F_2 progenies from an intra-species cross were used to produce a linkage map. An 80 Kb pepper BAC clone was found by BAC screening using a marker close to *Ptl1*. The BAC sequences were then used to develop a marker closer to *Ptl1*. Two markers were developed from the BAC clone sequence; one was mapped to a position 0.33 cM upstream from the *Ptl1* locus and the other to a position 0.75 cM downstream from it. This study opened the way for map-based cloning of *Ptl1* using pepper BAC clones.

5. The Comparative Analysis of the Pepper Genome Using Conserved Orthologous Set (COS) Markers

The pepper EST data enabled the generation of a genetic map using COS markers. These markers were developed using various existing plant EST databases, including those for pepper, tomato, tobacco and *Arabidopsis*. Comparative analysis of the EST databases allowed the identification of conserved orthologous EST sets, which were developed into the molecular markers used to generate the genetic map. By linking the COS markers in the genetic map, it became possible to compare the genomic structures between different species. Using this method, Wu et al. (2009) compared the pepper genome to that of the tomato. A total of 299 orthologous markers were used in the comparison, which included all of the 12 linkage groups (Chapter 3, Fig. 6-2). The analysis identified 19 inversions and six chromosomal translocations between pepper and tomato. Both genomes shared 35 conserved syntenic segments, indicating a highly conserved gene order and content, except for the inversions and translocations.

6. Ribosomal Repeats in the Pepper Genome

Several studies have attempted to distinguish the karyotypes of peppers. Possibly because of the large size of the genome, the pepper pachytene chromosomes are prone to lumping in the heterochromatic regions. Because of this, it is difficult to clearly distinguish different chromosomes from one another. Moreover, each pepper chromosome has a similar structure, and size differences between many of the chromosomes are very slight. Hence, distinction of the 12 chromosomes in the pepper genome remains to be accomplished. Karyotyping of peppers has been tried by fluorescence *in situ* hybridization (FISH) analyses using ribosomal genes. This method was used to study the differences between the karyotypes of several species of peppers. Tanksley et al. (1988) compared the karyotypes of *C. annuum* and *C. chinense* by FISH analysis using 45S rDNA, and the result revealed different ribosomal loci. Whereas the *C. annuum* karyotype showed only one ribosomal locus, the *C. chinense* karyotype showed two. Park et al. (1999) used 5S rDNA and 18S-26S rDNA as FISH probes to compare the karyotypes of *C. annuum*, *C. chinense*, *C. frutescens*, *C. baccatum* and *C. pubescens*. The results showed that the 5S rDNA appeared at only one site on chromosome 1 in all five species. However, the number and location of 18S–26S rDNA was highly variable among the five species (Fig. 6-2). Kwon et al. (2009) used 5S and 45S rDNA as FISH probes to compare the karyotypes of *C. annuum*, *C. chinense*, *C.frutescens*, *C.baccatum* and *C. chacoense*. This analysis gave the same result as that of Tanksley et al. for the 5S rDNA. The number of 45S rDNA loci varied among the five species. *C. annuum*

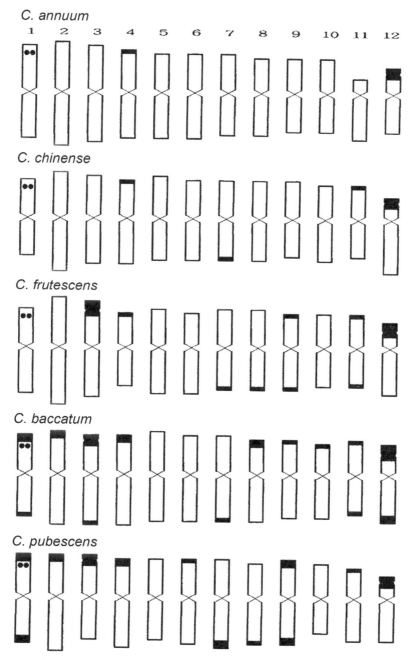

Figure 6-2 Locations of 18S-26S rDNA in the karyotypes of different *Capsicum* species.

contained only one locus; except for the CM334 variety. Although CM334 is classified as *C. annuum*, three 45S rDNA loci were identified in this variety. *C. chacoense, C. frutescens* and *C. chinense* each contained two 45S rDNA loci and *C. baccatum* contained four.

7. The Structure of the Pepper Genome

The structure of the pepper genome was studied by comparison with Solanaceous species such as tomato (Park et al. 2011). Compared to the tomato genome, the pepper genome is approximately three times larger and has a different chromosome structure. These differences were identified by comparing the pachytene chromosomes (Fig. 6-3). The DAPI-stained pachytene chromosome clearly reveals euchromatin and heterochromatin structure. The brightly and darkly stained regions indicate the regions of heterochromatin and euchromatin, respectively. The dramatic difference between pepper and tomato in chromosome size can also be seen. The composition of the euchromatin and the heterochromatin also differs between the two species. Whereas the boundary between the euchromatin and the heterochromatin is quite distinct in tomato (Fig. 6-3A), it is much less in pepper (Fig. 6-3B). In addition, the tomato heterochromatin is centered on the pericentromeric blocks, and the pepper heterochromatin is highly extended and is also found in the euchromatic regions (Fig. 3B; indicated by arrows).

Micro-synteny in pepper and other Solanaceous species was investigated by sequencing BAC clones of pepper, tomato, potato, eggplant and petunia containing a conserved syntenic segment (CSS) (Wang et al. 2008). Regions of approximately 100 Kb were compared with one another (Fig. 6-4). This

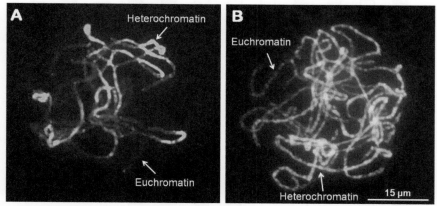

Figure 6-3 Pachytene chromosomes of tomato (A) and pepper (B). Brightly stained regions are heterochromatin and darkly stained regions are euchromatin.

study revealed conserved synteny and gene content at the sequence level, which was expected on the basis of the comparative analysis of the genetic maps generated by COS markers. This study allowed the estimation of the time of speciation between pepper and tomato, 19.2 million years ago.

Although the comparative analysis of the CSSs revealed highly conserved gene synteny, it did not explain the structural differences between the pepper and tomato genomes. This was closely examined in the study by Park et al. (2011), which investigated the cause of genome expansion in pepper by comparing orthologous BAC sequence pairs between pepper and tomato. This study used eight sequence pairs from six different loci on chromosome 2 (Fig. 6-5). The total size of the orthologous sequence pairs was 985,237 bp in pepper and 490,745 bp in tomato (Table 6-4). In this study, 94 percent of the total identified genes exhibited orthologous matches; the remaining 6 percent did not. The total number of identified genes was 145, and there were five more genes among the pepper sequences than among the tomato sequences. Approximately 25 percent of the genes were duplicates. The average lengths of the coding regions were similar between the two species, however, the average intron length was 356 bp longer in pepper. Although the gene synteny was well conserved, the gene density in pepper

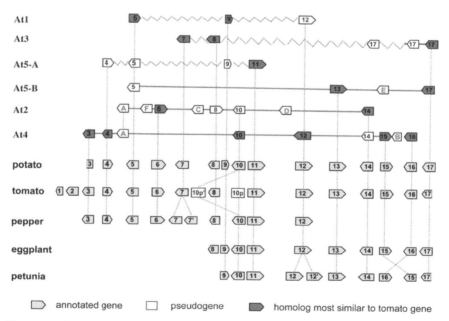

Figure 6-4 Comparative analysis of conserved syntenic segments in Solanaceae. The genes are depicted by arrows indicating the direction of transcription. The orthologous genes are connected by the dotted red lines.

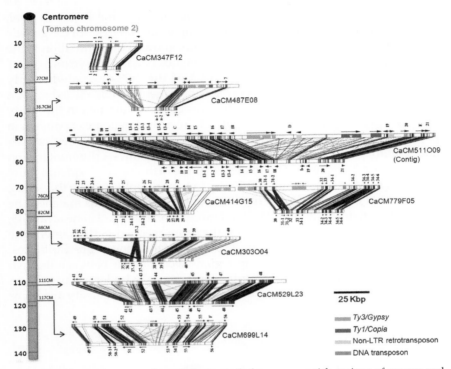

Figure 6-5 Sequence comparisons between orthologous gene-rich regions of pepper and tomato. The arrows indicate the predicted genes. The accession numbers indicate the orthologous gene sets. Colored bars indicate the repeat elements. The vertical lines indicate the highly similar regions.

Color image of this figure appears in the color plate section at the end of the book.

Table 6-4 Statistics of the pepper and tomato gene-rich sequences.

	Pepper	Tomato	Total (ratio)
Total length of compared sequences	985,237 bp	490,745 bp	-
Number of predicted genes	75	70	145
Total length of predicted genes	247,338 bp	195,342 bp	-
Gene density	13,136 bp/gene	7,011 bp/gene	-
Genes paired into orthologous sets	69	67	136 (94%)
Genes with no ortholog	6	3	9 (6%)
Duplicated genes	18	19	37 (25%)
Average length of coding region	1,366 bp	1,332 bp	-
Average length of intron	1,815 bp	1,459 bp	-

was approximately two times lower than that in tomato, indicating that the gene-rich regions in the pepper are expanded.

Comparative analysis revealed that the major reason for the euchromatin expansion in the pepper was the insertion of transposable elements (Fig. 6-5; colored boxes). All of the transposable elements were found in the intergenic regions of the pepper sequences without any disruption of the existing genes. The insertion of the transposable elements was identified in all of the pepper sequences, but was completely absent in the tomato sequences. The insertions resulted in a doubling of the total length of the pepper sequences.

Among the identified transposable elements, long-terminal repeat (LTR) retrotransposons were the most abundant (Fig. 6-6). Most of the LTR retrotransposons were found in the pepper sequences. The pepper sequences contained approximately 22 times more LTR retrotransposons than the tomato sequences. The other two repeat classes were also more abundant in pepper than in tomato. Pepper had about 1.7 times as many DNA transposons and four times as many non-LTR retrotransposons as the tomato had (Fig. 6-6). Among the repeat sequences found in pepper, approximately 64 percent were *Ty3/Gypsy*-like elements, suggesting an important role for these elements in the expansion of the pepper euchromatin.

A phylogenetic tree of the *Ty3/Gypsy*-like elements was generated to investigate their effects on the structure of the pepper genome. The sequences used to generate the phylogenetic tree were identified and obtained from the BAC sequence databases for pepper and tomato. Using

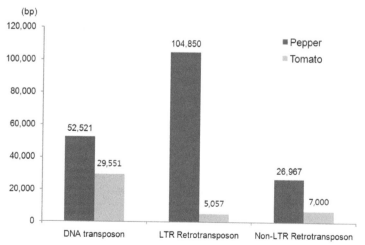

Figure 6-6 The numbers of repeat elements within the selected pepper and tomato sequences.

the phylogenetic tree, it was possible to compare the composition of the *Ty3/Gypsy*-like elements in pepper and tomato genomes (Fig. 6-7). The distribution of the elements within each subgroup was determined by

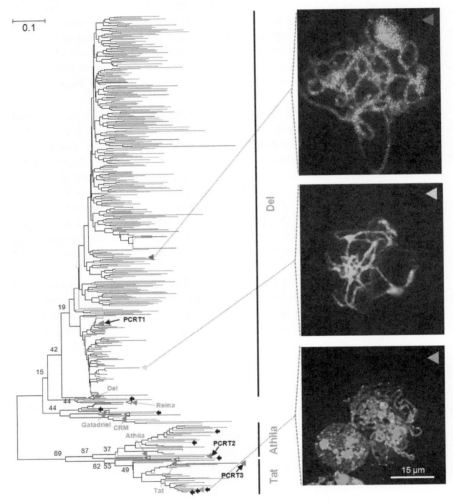

Figure 6-7 Phylogenetic analysis of pepper and tomato *Ty3/Gypsy*-like elements. The pepper and tomato *Ty3/Gypsy*-like elements are depicted by red and blue lines, respectively. The major subgroups Tat, Athila and Del are indicated by green letters. The reverse transcriptases (RTs) used as FISH probes are marked with triangles (purple, yellow and green). The FISH result for each of the probes is indicated by a dotted line. The black arrows indicate the RTs found in the selected gene-rich pepper sequences. The empty black triangles indicate the representative RTs for the different subgroups acquired from the GyDB.

Color image of this figure appears in the color plate section at the end of the book.

FISH analysis. The phylogenetic tree was largely divided into three major subgroups: Tat, Athila and Del. The FISH result for a Del element in pepper shows preferential distribution in the heterochromatin, indicating the accumulation of Del elements in heterochromatic regions (Fig. 6-7; indicated by purple triangle). A similar FISH result for a Del element in the tomato shows the same preferential distribution in heterochromatic regions (Fig. 7; indicated by yellow triangle). In another study using FISH analysis, the PCRT1 retrotransposon in the Del subgroup of tomato was also reported to be preferentially distributed in heterochromatic regions (Yang et al. 2005). Thus, the evidence suggests that pepper heterochromatin expanded via the accumulation of Del elements.

The cause of the expansion of pepper euchromatin was found in the Tat and Athila subgroups. The number of Tat and Athila elements found in pepper was approximately twice that found in tomato (42 in the pepper and 23 in the tomato). According to the report by Yang et al. (2005), the *Ty3/Gypsy*-like elements PCRT2 and PCRT3 are preferentially distributed in the heterochromatic regions of the tomato. These two elements belonged to the Athila and Tat subgroups, respectively, suggesting that these subgroups accumulate in the heterochromatic regions of the tomato. In contrast, the FISH result for pepper Tat subgroup showed randomly distributed signals throughout the pepper chromosomes, including the euchromatic regions (Fig. 6-7; indicated by green triangle). Furthermore, four of the nine black arrows, indicating the *Ty3/Gypsy*-like elements found in the gene-rich sequences of pepper belong to the Tat subgroup and two belong to the Athila subgroup. Hence, the Tat and Athila elements are relatively common in the gene-rich regions of the pepper. In contrast to their distribution in tomato, the *Ty3/Gypsy*-like elements in the Tat and Athila subgroups are distributed randomly throughout the whole genome of pepper, resulting in euchromatin expansion.

8. The Effort to Sequence the Whole Pepper Genome

The pepper whole genome sequencing project was launched in 2009 by the research group of Seoul National University in Korea. The whole genome and transcriptome of *C. annuum*, CM334 are currently being sequenced. The major strategy for the project involves using the Illumina Solexa GA system to produce massive amounts of raw sequences. To date (December, 2011), a total of 306 Gb of raw sequences, representing greater than 100× coverage, from the CM334 genome have been generated. The raw sequences have been successfully assembled and the results are as follows. The total length of the assembled contigs is 2,634 Mb and the total number of contigs is 1,858,508. The N50 size is 4,422 bp and the average contig length is 1,417 bp. To further extend the contigs, scaffold assembly is currently in progress.

In addition to the whole genome shotgun sequencing, pepper BAC clones have also been sequenced to make longer contigs. A total of 2,382 pepper BAC clones were selected by BAC library screening and used in the sequencing. A total of 1,270 BAC clones were sequenced using the Roche 454 GS-FLX sequencer and the others were sequenced by BAC pooling using the Illumina Solexa GA system. The BAC clones were screened by labeled pepper and tomato mRNA (tomato mRNA was used to avoid screening by retrotransposons). In addition, a total of 435 BAC clones containing genes belonging to the nucleotide-binding site-leucine-rich repeat (NBS-LRR) family were isolated by BAC screening using the NBS-LRR gene family sequences as probes. Recently, raw sequences of the 1,270 pepper BAC clones were assembled, resulting in 34,743 contigs with an average size of 2,707 bp and a total length of 0.94 Gb.

9. Future Prospects

The recent reductions in the costs of whole genome sequencing are expected to lead to further cost reductions in the future. This is transforming the pepper research environment. Southern blot analyses are being replaced by nucleotide BLAST searches against sequence databases to identify the copy numbers of specific genes; microarray techniques are being replaced by digital expression analyses of transcriptome sequencing data. The development of molecular markers can also be replaced by single-nucleotide polymorphism detection by re-sequencing. Recently, a project aimed at drawing a genetic map of the pepper by re-sequencing recombinant inbred pepper lines was begun, and the related works are in progress. The genes for specific phenotypes can be identified by association studies using re-sequencing data from diverse wild races of pepper. Although the pepper genome project is currently still generating the *de novo* reference genome sequence of *C. annuum*, CM334, this stage of the project is expected to be completed within a short time. The next goal will be re-sequencing of diverse *Capsicum* species. These sequence data will be used in diverse research areas, such as gene characterization, the development of molecular markers, genomic study and breeding.

References

Belletti P, Marzachì C, Lanteri S (1998) Flow cytometric measurement of nuclear DNA content in *Capsicum* (Solanaceae). Plant Syst Evol 209: 85–91.
Kim HJ, Baek KH, Lee SW, Kim J, Lee BW, Cho HS, Kim WT, Choi D, Hur CG (2008a) Pepper EST database: comprehensive in silico tool for analyzing the chili pepper (*Capsicum annuum*) transcriptome. BMC Plant Biol 8: 101.

Kim HJ, Han JH, Kwon JK, Park M, Kim BD, Choi D (2008b) Fine mapping of pepper trichome locus 1 controlling trichome formation in Capsicum annuum L. CM334. Theor Appl Genet 120: 1099–1106.

Kim HJ, Nahm SH, Lee HR, Yoon GB, Kim KT, Kang BC, Choi D, Kweon OY, Cho MC, Kwon JK, Han JH, Kim JH, Park M, Ahn JH, Choi SH, Her NH, Sung JH, Kim BD (2008c) BAC-derived markers converted from RFLP linked to *Phytophthora capsici* resistance in pepper (*Capsicum annuum* L.). Theor Appl Genet 118: 15–27.

Kwon JK, Kim BD (2009) Localization of 5S and 25S rRNA genes on Somatic and meiotic chromosomes in *Capsicum* species of chili pepper. Mol Cells 27: 205–209.

Moscone EA, Barany M, Ebert I, Greilhuber J, Eherendorfer F, Hunziker AT (2003) Analysis of Nuclear DNA Content in Capsicum (Solanaceae) by Flow Cytometry and Feulgen Densitometry. Ann Bot 92: 21–29.

Park M, Jo S, Kwon JK, Park J, Ahn JH, Kim S, Lee YH, Yang TJ, Hur CG, Kang BC, Kim BD, Choi D (2011) Comparative analysis of pepper and tomato reveals euchromatin expansion of pepper genome caused by differential accumulation of Ty3/Gypsy-like elements. BMC Genomics 12: 85.

Park Y-K, Kim B-D, Kim B-S, Armstrong KC, Kim N-S (1999) Karyotyping of the chromosomes and physical mapping of the 5S rRNA and 18S-26S rRNA gene families in five different species in *Capsicum*. Genes Genet Syst 74: 149–157.

Tanksley SD, Bernatzky R, Lapitan NL, Prince JP (1988) Conservation of gene repertoire but not gene order in pepper and tomato. Proc Natl Acad Sci USA 85: 6419–6423.

Tomita M, Noguchi T, Kawahara T (2009) Quantitative variation of Revolver transposon-like genes in synthetic wheat and their structural relationship with the LARD element. Breed Sci 59: 629–636.

Wang Y, Diehl A, Wu F, Vrebalov J, Giovannoni J, Siepel A, Tanksley SD (2008) Sequencing and comparative analysis of a conserved syntenic segment in the Solanaceae. Genetics 180: 391–408.

Wu F, Eannetta NT, Xu Y, Durrett R, Mazourek M, Jahn MM, Tanksley SD (2009) A COSII genetic map of the pepper genome provides a detailed picture of synteny with tomato and new insights into recent chromosome evolution in the genus *Capsicum*. Theor Appl Genet 118: 1279–1293.

Yang TJ, Lee S, Chang SB, Yu Y, de Jong H, Wing RA (2005) In-depth sequence analysis of the tomato chromosome 12 centromeric region: identification of a large CAA block and characterization of pericentromere retrotranposons. Chromosoma 114: 103–117.

Yoo EY, Kim YH, Lee CJ, Kim BD (2003) Construction of a deep coverage BAC library from *Capsicum annuum*, 'CM334'. Theor Appl Genet 107: 540–543.

Eggplant

Amy Frary[1] and Sami Doganlar[2,]*

ABSTRACT

Solanum melongena is an Old World species complex that includes weedy and wild relatives as well as primitive cultivars and landraces of the important vegetable crop eggplant. While the origins of cultivated eggplant are obscure, most evidence suggests an Indo-Chinese centre of domestication. Sexual promiscuity within eggplant and its relatives blurs species boundaries, making taxonomic relationships difficult to decipher. Furthermore, attempts to define finer-scale evolutionary relationships are thwarted by low levels of genetic variation (despite considerable morphological diversity) in cultivated eggplant. The main breeding objectives in the crop are to increase yield by heterosis breeding, introduce disease and pest resistances from wild germplasm, and improve fruit quality through selection for parthenocarpy and elevated levels of key secondary metabolites (anthocyanins and phenolics). Extensive germplasm resources collected in Asia and Europe hold considerable potential for the genetic improvement of the crop; however, the introgression of traits from wild relatives is hampered by low fertility in hybrids. A variety of molecular markers have been used to characterize the germplasm collections. This work has helped both to classify species and to identify potentially valuable sources of heterozygosity for modern cultivars. Molecular markers have also been used to construct linkage maps of the eggplant genome. The close relationship between eggplant, tomato and pepper has facilitated this work as well as made the Solanaceae a model for comparative genomics. The molecular genetic maps developed in eggplant have been used both for the tagging of simply inherited traits and the localization of the loci underlying complex morphological characters. Quantitative trait analysis of interspecific mapping populations indicates that

[1]Mount Holyoke College, South Hadley, Massachusetts, USA.
[2]Izmir Institute of Technology, Izmir, Turkey.
*Corresponding author

wild germplasm holds promise for improving fruit yield and quality. These analyses also suggest that the conservation of gene function and position between the tomato and eggplant genomes should allow the knowledge gained in tomato to be leveraged for the improvement of eggplant. The development of an integrated molecular linkage map, generation of expressed sequence data and a growing interest in the potential nutritional and medicinal benefits of eggplant promise productive years ahead.

Keywords: Solanum melongena, Evolution, Germplasm, Breeding, Mapping

1. Economic Importance

Solanum melongena, known variously as eggplant, brinjal and aubergine, is an important vegetable crop throughout Asia and the Mediterranean region where its fruits are a key ingredient of national and regional cuisines. While worldwide production of this Old World solanaceous crop species lags behind that of its New World kin, potato and tomato, 43 million tons (t) of eggplant were harvested from 1.7 million hectares (Ha) of land in 2009 (FAO 2009). Asia is the principal centre of eggplant production. In 2009, China grew 26 million tons of the crop on 740,000 Ha and India 10 million tons on 600,000 Ha. While far less of the crop was grown in Egypt (1.2 million t; 50,000 Ha) and Turkey (820,000 t; 27,000 Ha), these two countries are ranked as the 3rd and 4th largest producers of eggplant worldwide. Other notable producers are Indonesia (450,000 t; 46,000 Ha), Iraq (396,000 t; 21,000 Ha), Japan (349,000 t; 10,400 Ha), Italy (245,000 t; 9400 Ha), Spain (205,000 t; est. 3500 Ha) and the Philippines (201,000 t; 21,000 Ha) (FAO 2009). The average yield (26 t/Ha) ranges considerably, depending on environmental and cultural conditions with the highest yields achieved in the Netherlands (460 t/Ha) from F_1 hybrids grafted onto disease resistant rootstocks and grown under highly controlled greenhouse conditions.

2. Nutritional Properties

While the fruit of the eggplant is used extensively in Asian, Indian and Mediterranean cookery, its nutritional value is modest (Gebhardt and Thomas 2002). Despite being a poor source of protein (1.0g/100g fresh weight), provitamin A (27 IU/100g) and vitamin E (0.30 mg/100g) (USDA 2011), eggplant is rich in some minerals and antioxidants. For example, 100 g of eggplant (fresh weight) can provide ~5 percent of the recommended daily amount (RDA) of phosphorus, potassium and copper as well as ~10 percent of the daily intake of phenolics (Raigón et al. 2008). Anthocyanins, another important class of antioxidants, are found in abundance in the richly pigmented peel of the fruit (Azuma et al. 2008). Extensive research into the health benefits provided by the free radical scavenging properties of both phenolics and anthocyanins has been conducted in recent years (Kaur and Kapoor 2001; Liu 2003). In addition, the alkaloids produced by eggplant and its wild relatives have been reported to induce apoptosis in tumor cells (Kuo et al. 2000; Cham 2007). Thus, while eggplant has been used in traditional Chinese and Indian medicine for centuries (Daunay and Janick 2007; Daunay et al. 2007), the biological bases of its merits as a medicinal plant have only recently been revealed.

3. Brief History of the Crop

The origins of cultivated eggplant are somewhat obscure. Phylogenetic studies have revealed that *Solanum incanum* is the closest relative of *S. melongena* (Lester and Hasan 1991). However, *S. incanum* itself is a species complex that encompasses up to 27 species and 81 taxa (Bitter 1923) endemic to eastern Africa and the Middle East. These taxa have been divided into four separate and diverse groups (A-D) of wild eggplant relatives on the basis of their distribution and morphological characters (Lester and Hasan 1991). It has been hypothesized that *S. incanum* group C arose as plants of groups A and B spread from Africa to the Middle East and became adapted to local conditions. Then, continuing its eastward migration, group C *S. incanum* reached Asia and diverged to produce *S. melongena*. Group D represents the types that were moved to South Africa and adapted to the arid conditions there (Daunay et al. 2001).

Like *S. incanum*, *S. melongena* is a species complex that comprises four distinct groups (E-H) (Lester and Hasan 1991; Mace et al. 1999). Group F (also known as *S. cumingii*) is the likely ancestor of domesticated eggplant and includes weedy forms found widely in Indochina and Indonesia. Group G (*S. ovigerum*) represents domesticated primitive cultivars and landraces with small round or oblong fruits which grow throughout Southeast Asia. Group H (*S. melongena*) includes the large-fruited advanced cultivars that are grown in agricultural fields and greenhouses worldwide. Evidence suggests that *S. melongena* Group E (*S. insanum*) arose as revertants from group G forms (Lester and Hasan 1991; Mace et al. 1999). This group includes weedy plants with wild attributes. While the direct evidence in support of this scenario of evolution and domestication in eggplant is currently scarce, an analysis of DNA sequences (ITS, *waxy* and *trnT-F*) from eggplant and its wild and weedy relatives indicates that the progenitors of eggplant arose in Africa (Weese and Bohs 2010).

Based on the current distribution of *S. melongena* group F, it is possible to deduce an Indo-Chinese centre of origin and domestication for eggplant. Moreover, because of eggplant's importance as a food, medicinal and ornamental plant to the ancient civilizations of this region, the spread and early history of the crop can been studied through the examination of historical records, literature and iconography (Daunay and Janick 2007; Daunay et al. 2007; Wang et al. 2008). Eggplant is frequently mentioned by various common names in Sanskrit documents, some of which date to 300 BCE (Daunay and Janick 2007). The earliest known written record of cultivated eggplant in Chinese literature dates to 59 BCE (Wang et al. 2008). Thus, eggplant is a crop that dates to antiquity. Around the 8th century, eggplant spread eastward to Japan and then westward to the Mediterranean (perhaps brought by Muslim conquerors) (Frary et al. 2007). Descriptions in Chinese literature (6th-18th centuries) as well as images of eggplant from Chinese and European herbals (dating to the 14th–17th centuries) depict the progressive changes in fruit sizes, shapes, and colors wrought by domestication and cultivation: from small round white fruit to larger fruit of various hues (white, green, violet and almost black) and shapes (round, oblong, long and thin) (Daunay and Janick 2007; Wang et al. 2008) (Fig. 7-1). Changes in the flavour of eggplant (from bitter to sweet) have also been gleaned from Chinese writings (Wang et al. 2008).

Figure 7-1 Depictions of eggplant in Chinese literature: (A) *Tu Jing Bencao* (AD 1069), (B) *Bencao Gangmu* (AD 1590), (C) *Sancai Tu Hui* (AD 1609), (D) *Zhiwu Ming Shi Tu Kao* (AD 1848). Reproduced by permission of the Chinese Academy of Sciences. (Used with permission from Wang et al. 2008, Oxford University Press.)

4. Taxonomy

S. melongena is a member of the Solanae tribe within the Solanaceae family. This tribe includes the New World crop plants: potato (*S. tuberosum*), tomato (*S. lycopersicum*) and pepper (*Capsicum* sp.). Eggplant is further classified as one of the "spiny solanums", *Solanum* subgenus *Leptostemonum*. This subgenus includes 350–450 species of New and Old World plants, typified by leaf and stem prickles (Levin et al. 2006). Two other species commonly referred to as eggplants, *Solanum macrocarpon* (gboma eggplant) and *S. aethiopicum* (scarlet eggplant), also belong to this group. Both are cultivated in Africa as fruit and vegetable crops. Taxonomic relationships in the subgenus and particularly within eggplant and its relatives are confusing; extensive morphological diversity and interspecific crossability blurs species boundaries and evolutionary relationships. Nevertheless, the Old World species reliably group as a clade in molecular phylogenies of the subgenus (Levin et al. 2006). Wild and weedy forms of *S. incanum* cluster separately from *S. melongena*, and scarlet and gboma eggplants fall outside of the *incanum-melongena* complex (Weese and Bohs 2010). The low levels of genetic variation combined with sexual promiscuity within the *S. melongena* complex make it difficult to determine finer-scale evolutionary relationships (Weese and Bohs 2010).

5. Botanical Description

In its primitive form, eggplant is a tall, woody perennial plant with large leaves. Prickles on the stem, leaves and calyx are typical. Andromonoecious flowers with five connate sepals, five connate petals and five stamens fused to the corolla are produced in small cymes (one to five flowers/inflorescence). Flowers generally self-pollinate although cross-pollination may occur in nature as a result of heterostyly and insect visitation. The fruit, berries, are small and thick-skinned, green and hard at maturity and unpalatable because they are bitter and seedy. Wild eggplants are distributed in tropical regions of Africa and Asia. Domestication, cultivation and breeding have resulted in a smaller plant that is grown as an annual crop worldwide. Eggplant is a field crop in the Middle East and much of Asia, but greenhouse production is on the rise, especially in Europe and Japan (Daunay 2008). Cultivated forms usually lack prickles and produce perfect flowers that are often solitary. Fruits have thin skin, soft flesh and are larger, less seedy and less bitter than wild types. A diversity of fruit shapes exists among cultivars, with round, ovate, oblong, fasciated, elongate and serpentine forms. Fruit size varies widely along with shape. Fruit length ranges from 4 to 45 cm and diameter from 2 to 35 cm. A 100-fold difference in fruit weight (15 g to 1.5 kg) is seen among varieties. Fruit colors are linked to the variable presence of chlorophyll (green) and anthocyanin (red and purple) pigments in the developing fruit. White, green, violet, purple and almost black varieties have been selected, some with contrasting stripes or streaks (Swarup 1995; Frary et al. 2007). Eggplant's genome consists of approximately 956 Mbp (Bennett and Leitch 2010); it is a diploid with a base chromosome number of 12.

6. Germplasm

Collections of eggplant cultivars, landraces, relatives and wild species in Asia include ~1,800 accessions in India (National Bureau of Plant Genetic

Resources, NBPGR, New Delhi) (Gangopadhyay et al. 2010), nearly 2000 accessions in China (Institutes of Vegetables Crops, IVC, Nanjing and Hangshu) (Mao et al. 2008) and ~400 lines in Japan (National Institute of Agrobiological Sciences, NIAS, Tsukuba). Some germplasm resources have also been collected in Southeast Asia (Wivutvongvana et al. 1984; Sakata et al. 1996), Indonesia (Gousset et al. 2005), Africa (Lester et al. 1990) and the Middle East (Sadder et al. 2007). A comprehensive database of eggplant-related germplasm holdings in Europe (estimated to include ~6,000 accessions) (Daunay et al. 2003) was compiled by the EGGNET (Eggplant Genetic Resources Network) project, a network of public and private sector researchers from Netherlands, France, Italy, Spain, Greece, Germany and UK. The database is currently curated by the European Cooperative Programme for Plant Genetic Resources (ECPGR), Nijmegen, Netherlands. The Germplasm Resources Information Network (GRIN) database at the USDA maintains over 800 lines of *S. melongena* and related species, including the African eggplants.

Tremendous morphological diversity is available in primitive cultivars, landraces and weedy species within *S. melongena*. In addition, the wild progenitor, *S. incanum* represents an important source of resistance to biotic and abiotic stresses (Swarup 1995). Useful germplasm for the genetic improvement of cultivated eggplant can also be found among wild allied species, many of which harbour disease and pest resistance genes (reviewed in Collonnier et al. 2001). Eggplant can be easily crossed with a wide variety of these spiny solanums. However, while 19 related species have been employed in the breeding of eggplant, only four of them yield fertile progeny when crossed with *S. melongena*: *S. incanum*, *S. linnaeanum*, *S. macrocarpon* and *S. aethiopicum* (Collonnier et al. 2001). Thus, in practical terms, the introgression of donor genes from more distantly related eggplant relatives is challenging.

7. Classical Genetics and Traditional Breeding

While *Solanum lycopersicum* (tomato) was an important model organism in early classical mapping efforts, *S. melongena* was essentially ignored in this regard. With the exception of anthocyanin accumulation (Nolla 1932; Janick and Topoleski 1963; Tigchelaar et al. 1968), very few phenotypic traits have been mapped in eggplant. The quantitative nature of many key agronomic characters has complicated inheritance studies in this as in other crop species. With the advent of molecular linkage maps and the concomitant development of comparative genomics, attention once focused exclusively on tomato has widened to encompass eggplant (as well as potato and pepper). Genome-wide characterization of eggplant through molecular mapping can facilitate breeding on a number of levels. The inheritance of complex traits can be more easily analyzed, unwanted genotypes can be culled from breeding populations earlier via marker-assisted selection and germplasm can be screened more efficiently for desirable traits (Frary et al. 2007).

The main objectives of eggplant breeders have been to increase yield to incorporate disease and pest resistance into the crop and to improve fruit quality. Improved tolerance to abiotic stress is another important goal. Since the discovery of heterosis in eggplant (Kakizaki 1931) breeding efforts have increasingly focused on developing hybrids with enhanced productivity from inbred materials. As a result, most commercial cultivars are F_1 hybrids. And yet, eggplant breeding is hampered by the labor-intensive aspects of hybrid seed production. Manual emasculation and pollination of the inbred parents is time-consuming and expensive. Therefore, efforts to introduce cytoplasmic male sterility (CMS) into breeding lines of eggplant are ongoing. Male sterile lines have been generated by interspecific crosses between *S. melongena* (as the male parent) and *S. violaceum* (Isshiki and Kawajiri 2002), *S. virginianum* (Khan and Isshiki 2008), *S. grandifolium* (Saito et al. 2009a) and *S. anguivi* (Khan and Isshiki 2010), followed by several generations of backcrossing to *S. melongena*. In all of these cases, the incorporation of the foreign species' cytoplasm had the desired effect of reducing pollen production or pollen fertility but leaving seed fertility unaltered. The successful integration of CMS in commercial hybrid seed production demands that the sterility be stable in a range of genetic backgrounds (Saito et al. 2009a) and that nuclear fertility restorer (*Rf*) genes be present. Such genes have been identified in progenies from crosses with *S. anguivi* (Khan and Isshiki 2010) and *S. grandifolium* (Saito et al. 2009a).

In addition to heterosis breeding, advances in eggplant productivity have been achieved by grafting *S. melongena* onto tomato (*S. lycopersicum* and *S. hirsutum*) as well as more closely related species (*S. torvum*, *S. aethiopicum*, *S. macrocarpon* and *S. incanum*) (Daunay 2008; Gisbert et al. 2011). Increased vigour of the scion as well as earlier fruit maturity are some of the benefits of grafting (Khah 2005; Gisbert et al. 2011). In addition, grafting is used as a means of conferring resistance to soil pathogens onto susceptible eggplant cultivars, thereby boosting yield (Bletsos et al. 2003; reviewed in Daunay 2008).

The diseases of greatest economic concern in eggplant are the wilts caused by *Ralstonia solanacearum* (bacterial wilt), *Fusarium oxysporum* f. sp. *melongenae*, and *Verticillium dahliae*. Fruit and shoot borer (*Leucinodes orbonalis*) is the most significant insect pest. Aphids (*Aphis gossypii*), nematodes (*Meloidogyne* spp.) and little leaf disease (caused by a mycoplasma) also impact yield (Daunay 2008). While disease and pest resistance levels are generally quite low in commercial varieties, tolerance has been identified in a number of wild relatives of eggplant (reviewed in

Collonnier et al. 2001), and efforts are ongoing to introgress resistance genes into cultivated germplasm. The success of such crosses depends in large part on the phylogenetic distance between the parents, the *S. melongena* genotype and the direction of the cross (Schaff et al. 1982; Bletsos et al. 2000). A major obstacle has been cross-species incompatibilities that appear as low fertility or sterility in the interspecific hybrids and are often attributed to meiotic irregularities (Schaff et al. 1982; Behera and Singh 2002; Bletsos et al. 2004). Chromosome doubling (achieved via colchicine treatment or anther culture) has been an effective means of restoring hybrid fertility in some cases (Isshiki et al. 2000; Isshiki and Taura 2003).

Somatic hybridization is another strategy that has been used to create interspecific eggplant hybrids (Kameya et al. 1990; Jarl et al. 1999; Collonnier et al. 2003; Iwamoto et al. 2007). Although such hybrids typically express the desired trait they also have a tendency toward sterility. While exceptions to this have been reported (Borgato et al. 2007; Iwamoto et al. 2007), the integration of somatic hybrids into breeding programs is further hampered by their tetraploid nature. Anther culture, however, has proven to be useful for bringing such lines back to diploidy (Rizza et al. 2002; Rotino et al. 2005). Regardless of the trait of interest, when trying to introgress genes from wild species into cultivated germplasm, it is critical that meiotic recombination occurs between homeologues of the two parental species. Evidence for such chromosomal exchange has been found in *S. melongena + S. aethiopicum* somatic hybrids (Rizza et al 2002; Rotino et al. 2005; Toppino et al. 2008a). Such hybrids could be employed to transfer bacterial and Fusarium wilt resistances into eggplant.

Fruit quality traits include the flavour and texture of the flesh, the thickness, color and glossiness of the skin and the storability of the harvested fruit. Recently, there has been an emphasis on breeding for parthenocarpy (the development of seedless fruits) (Fig. 7-2). Parthenocarpy ensures fruit set and development under otherwise adverse environmental conditions, such as sub- or supra-optimal temperatures and humidity levels, low light and heavy rain or strong wind. In addition, seedless fruits typically are less bitter, have softer flesh and brown less quickly than their seeded counterparts (Donzella et al. 2000). Treating flowering plants with phytohormones can induce parthenocarpy, however the labor and expense associated with such treatments can be prohibitive (Kikuchi et al. 2008a). A more sustainable approach is to select for the trait, and parthenocarpic cultivars have been developed in this way (Kikuchi et al. 2008b; Saito et al. 2009b).

Anthocyanin pigments are significant determinants of fruit quality in eggplant because of their obvious effects on color as well as their antioxidant activity and potential health benefits (Noda et al. 2000; Sadilova et al. 2006; Nisha et al. 2009). An increased interest in plant secondary metabolites has thus shifted attention towards breeding for these pigments and other

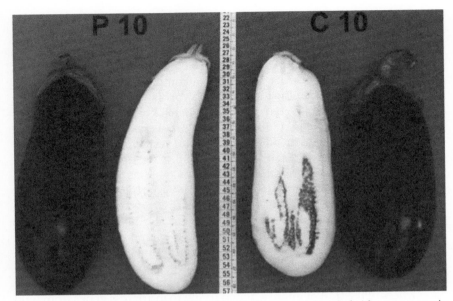

Figure 7-2 The exterior and interior of fruit of two eggplant hybrids: a transgenic parthenocarpic line containing the *iaaM* gene (P10) and a control (C10). (Used with permission from Acciarri et al. 2002, licensee BioMed Central Ltd.)

Color image of this figure appears in the color plate section at the end of the book.

compounds that impact nutritional quality. Several recent studies have focused on determining the antioxidant content in eggplant cultivars, landraces and related species (Azuma et al. 2008; Mennella et al. 2010). Thus, the anthocyanin profiles of numerous accessions of eggplant and related species have been compared; and, based on the radical-scavenging activities of purified pigments, it has been suggested that lines containing the anthocyanin delphinidin 3-glucoside should have the highest antioxidant properties (Azuma et al. 2008).

Over 14 different phenolic compounds, another important class of antioxidant, have been identified in eggplant accessions (Stommel and Whitaker 2003). Researchers in Taiwan, Spain, Turkey and Italy have measured total phenolics in a number of varieties and hybrids (Hanson et al. 2006; Raigón et al. 2008; Okmen et al. 2009; Mennella et al. 2010). The results of these studies indicate that the levels of phenolics in the different lines are fairly variable (~2-fold variation was generally observed in the studies) and that it should be possible to select materials with enhanced antioxidant capacity for breeding purposes. Because phenolics contribute toward oxidative browning of cut fruit, a negative quality trait, a tradeoff between these two attributes would seem to be necessary. However, heritability studies in over 100 eggplant varieties and landraces suggest that only around 20 percent of the variability in degree of browning can be attributed to phenolics levels, suggesting that it should be possible to obtain lines with high phenolics content but acceptable levels of oxidative browning (Prohens et al. 2007, 2008).

Other secondary metabolites of interest in eggplant are the glycoalkaloids solamargine and solasonine, which are present in all three cultivated species, *S. melongena, S. aethiopicum* and *S. macrocarpon* (Mennella et al. 2010; Sánchez-Mata et al. 2010). Solamargine has been investigated as a possible anti-cancer agent (Kuo et al. 2000; Shiu et al. 2007), and both compounds appear to be effective in killing parasitic trypanosomatids (Hall et al. 2006). Within the S. melongena germplasm tested, pickling varieties held the highest concentration of solamargine (Sánchez-Mata et al. 2010). However, the levels of these two glycoalkaloids in some allied species (namely, *S. macrocarpon, S. sodomaeum, S. aethiopicum* and *S. integrifolium*) may be high enough to warrant the concern about potential toxicity should these species be used in *S. melongena* breeding programs (Mennella et al. 2010; Sánchez-Mata et al. 2010). Further assessment of the safety of these compounds is necessary before they become a breeding priority.

8. Genetic Engineering

Eggplant has proven very amenable to genetic engineering; protocols for the regeneration of plants from *in vitro* cell and tissue cultures, somatic hybridization and *Agrobacterium*-mediated transformation are well established in the crop (reviewed in Rajam and Kumar 2007). As described previously, somatic hybridization has been a particularly useful tool in transferring disease resistances into eggplant from wild relatives. Insect resistance, on the other hand, has been the primary goal of transformation experiments in *S. melongena. Cry* genes coding for the insecticidal crystal protein endotoxin from *Bacillus thuringiensis* have been inserted into the eggplant genome via *Agrobacterium*-mediated transformation in several experiments. The resulting regenerants have shown resistance to Colorado potato beetle (*Leptinotarsa decemlineata*) (Arpaia et al. 1997; Hamilton et al. 1997) and fruit and shoot borer (*Leucinodes orbonalis*) (Kumar et al. 1998; Pal et al. 2009). However, in India, where the fruit and shot borer causes extensive yield losses, attempts to introduce commercial varieties of transgenic eggplant carrying the *Cry1Ac* gene have thus far been unsuccessful (Shelton 2010). Aphid resistance in eggplant has also been achieved through insertion of the rice oryzacystatin gene (Ribeiro et al. 2006), and the tomato *Mi-1.2* gene was found to increase eggplant's tolerance to the root-knot nematode species *Meloidogyne javanica* (Goggin et al. 2006).

Parthenocarpic eggplant lines have been engineered through the introduction of the *iaaM* gene from *Pseudomonas syringae* pv. *savastanoi* (Rotino et al. 1997; Acciarri et al 2002). This gene plays a key role in the biosynthesis of auxin, the plant hormone that is used to induce parthenocarpy in eggplant when environmental conditions are unfavorable for fruit set and development. In practice, treatment of plants with phytohormones is laborious, time-consuming and expensive. Thus, the potential economic benefits of transgene-induced parthenocarpy are considerable. In addition to the enhanced quality expected of seedless fruit, genetically-modified (GM) eggplants gave higher yields in both field and greenhouse (Donzella et al. 2000; Acciarri et al. 2002). Elevated levels of tolerance to salt, drought and cold temperature stresses were observed in genetically-modified eggplants carrying the mannitol-1-phospho dehydrogenase (*mtlD*) gene (Prabhavati et al. 2002). Although several noteworthy successes have been realized in the genetic engineering of eggplant, consumer wariness of GM vegetable crops is such that commercial growers have not yet adopted transgenic varieties.

9. Diversity Analysis

The morphological, agronomic and molecular diversity of eggplant and allied species has been examined by a number of researchers in efforts to assess genetic relatedness to characterize germplasm collections and to guide breeding programs towards promising lines. In most of the morphological studies, the vegetative, floral and fruit traits that were analyzed correspond to the primary characterization descriptors developed by EGGNET. A variety of molecular markers have been utilized, including allozymes, chloroplast DNA, random amplified polymorphic DNA (RAPD), simple sequence repeat (SSR), inter-simple sequence repeat (ISSR), amplified fragment length polymorphism (AFLP) and sequence-related amplified polymorphism (SRAP) markers.

The advantages of molecular traits over morphological ones in assessing both diversity and relatedness in eggplant have been highlighted by a number of studies. A study of 16 descriptors in 98 accessions of *S. melongena* and the African eggplants, *S. aethiopicum* and *S. macrocarpon*, demonstrated that considerable morphological diversity exists both within and between species (Polignano et al. 2010). Cluster analysis of the phenotypic data resulted in three groups that were unrelated to the classification of an accession (subspecies, botanical or variety group, cultivar or population), thus suggesting that morphology is an unreliable predictor of genetic status. The authors, therefore, suggest that molecular markers are an essential means of categorizing germplasm collections. Allozyme and RAPD data suggest that while *S. melongena*, *S. insanum* (used to denote weedy forms) and *S. incanum* (the wild progenitor of cultivated eggplant) are morphologically distinct, they are very similar at the genetic level (Karihaloo and Gottlieb 1995; Karihaloo et al. 1995). Even though weedy eggplants show more diversity than the cultivated accessions, the authors suggest that separate species designations are meaningless. Similarly, restriction fragment length polymorphism (RFLP) analysis of chloroplast DNA from *S. melongena* and related species indicated that taxonomic relationships as deduced from morphological characters are somewhat unreliable (Sakata and Lester 1997).

However, similar dendrograms have been obtained from both molecular data (AFLP and RAPD markers) and phenotypic data in comparative analyses of *Solanum* accessions (eggplant cultivars, landraces and related species) (Furini and Wunder 2004; Sadder et al. 2007). In addition, both morphological and molecular data were seen to be useful in the correct classification of previously unnamed and mis-named lines (Furini and Wunder 2004), suggesting that morphological traits still have a place in phylogenetic analysis of this taxonomically confusing group of plants.

Obviously, the choice of markers and accessions will have a large impact on the conclusions drawn from molecular analyses of diversity. Thus, while low genetic diversity was detected within *S. melongena* with microsatellite markers (Nunome et al. 2003a,b), other markers, including RAPD (Singh et al. 2006), SSR (Behera et al. 2006), genic SSRs (Stàgel et al. 2008; Tümbilen et al. 2011a) and SRAP (Li et al. 2010), have been generated, which reveals substantial variation both within *S. melongena* and among allied species. Some of these markers are sufficiently variable to reveal phylogenetic relationships (Furini and Wunder 2004; Behera et al. 2006; Stàgel et al. 2008; Isshiki et al. 2008; Li et al. 2010). Highly discriminatory ISSR and RAPD markers which are useful for fingerprinting cultivars have also been developed (Isshiki et al. 2008; Tiwari et al. 2009).

Regardless of the marker type used, it is clear that although morphologically diverse, cultivated eggplant has a much narrower genetic background than related allied species, including *S. incanum*, the aggregate of species that includes the progenitor of *S. melongena* (Nunome et al. 2003a; Tümbilen et al. 2011a). This low diversity has been attributed to a founder effect that likely occurred during the migration of *S. incanum* from Africa to Asia as well as the genetic bottleneck within *S. melongena* that probably accompanied the transition to the domesticated form. A comparison of variability in commercial cultivars (both F_1 and nonhybrids) and landraces within black-fruited accessions of *S. melongena* uncovered much higher levels of phenotypic and genotypic (based on polymorphism in SSR markers) diversity among landraces than commercial materials (Muñoz-Falcón et al. 2008). Moreover, the F_1 hybrids were revealed to share a very limited gene pool. Therefore, although domestication and breeding efforts have served to constrain diversity within eggplant, landraces and non-hybrid varieties are potentially valuable sources of heterozygosity for modern cultivars.

Thus, pockets of diversity can be found within local varieties of eggplant. Comparing traditional Spanish cultivars to four control accessions on the basis of morphology and AFLP markers, Prohens et al. (2005) discovered that "round" cultivars showed greater genetic diversity than "semi-long", "long" and "listada de Gandía" types (the four classes of local varieties). Sufficient levels of overall variability were found to suggest that Spain is a secondary centre of diversity for eggplant. A separate examination of "listada de Gandía" (a well-known heirloom) and other "listada" (striped) accessions from Spain and elsewhere revealed that the Spanish "listada" lines could be distinguished by their spinier calyces and heavier fruit as well as slower reproductive development (Muñoz-Falcón et al. 2008). And while "listada de Gandía" showed low levels of genetic diversity, the absence of three AFLP fragments in those accessions represent a unique genetic fingerprint that could enable legal protection of this heirloom variety.

Turkey is also considered a secondary centre of eggplant diversity, given the wide range of types cultivated in that country. Varieties are designated as "round", "semi-long", and "long", and these forms are used differently in preparing Turkish cuisine. Morphological and molecular AFLP analysis in 67 eggplant lines from the Turkish national germplasm collection showed the greatest phenotypic variability in "round" and "semi-long" forms. While these different morphological types did not form separate clusters on the basis of the molecular data, genetic similarity levels in the accessions (0.30 to 0.95) were indicative of intermediate diversity (Tümbilen et al. 2011b). The value of preserving local germplasm as well as the advantages of molecular tools in the efficient classification and management of such germplasm are made apparent by these studies (Prohens et al. 2005; Muñoz-Falcón et al. 2008; Tümbilen et al. 2011b). Genotypic data from molecular markers can be easily obtained within a couple of weeks of seed germination whereas phenotypic evaluation of vegetative and reproductive traits can take a month or more. Once generated, such markers can also be integrated into gene mapping studies.

10. Molecular Linkage Maps

10.1 A History

As previously mentioned, initial mapping efforts in the Solanaceae focused on the two most important crops in the family, tomato (*S. lycopersicum*) and potato (*S. tuberosum*) (Tanksley et al. 1992). The first linkage map for eggplant was developed on the basis of a population of 168 F$_2$ individuals generated from a cross between *S. melongena* breeding lines from Japan (EPL1) and

India (WCGR112-8) (Nunome et al. 1998). Segregation analysis of 94 RAPD markers yielded a map encompassing 716.7 cM and 13 linkage groups with an average marker spacing of 8.8 cM. The subsequent characterization of AFLP markers in this population [using MapMaker/EXP 3.0 (Lander et al. 1987)] improved the density of the map; 191 markers (88 RAPD and 93 AFLP) spanning 779.2 cM were spaced on average 4.9 cM apart on 21 linkage groups (Nunome et al. 2001). However, because the number of linkage groups far exceeded eggplant's chromosome number of 12, the fragmentary and incomplete nature of the map was obvious. The addition of seven polymorphic SSR markers brought the linkage groups down to 17 (Nunome et al. 2003b). More recently, a large number of SSRs were characterized in this intraspecific population and a 959.1 cM map consisting of 14 linkage groups and 236 SSR markers at an average spacing of 4.3 cM was generated using Mapmaker/EXP 3.0 (Lander et al. 1987; Nunome et al. 2009). Clustering of the genomic SSR markers in heterochromatic chromosomal regions was observed, as expected, given the unequal distribution of microsatellites within the genome (Broun and Tanksley 1996).

An interspecific cross, *S. linnaeanum* MM195 x *S. melongena* MM738, was used to create a molecular map of the eggplant genome based on tomato markers (Doganlar et al. 2002a). Thus, 58 F_2 progeny were genotyped, and a map comprising 233 RFLP markers and spanning 1480 cM and 12 linkage groups was constructed using Mapmaker v. 2.0 software (Lander et al. 1987). This map had the distinct advantage of allowing direct comparisons to be made between the eggplant and tomato genomes (discussed later). Using JoinMap v. 1.3 (Stam 1993) Sunseri et al. (2003) also generated a linkage map from an interspecific *S. sodomeum* (= *S. linneanum*) × *S. melongena* F_2 population (48 individuals). The map is based on segregation data for 117 RAPD and 156 AFLP markers that range over 736 cM on 12 linkage groups, with an average marker interval of 2.7 cM.

More recently, a high-resolution synteny map of eggplant was constructed by placing 110 COSII markers on the Doganlar et al. (2002a) map (Wu et al. 2009a) (Fig. 7-3). The COSII markers were developed from single copy genes conserved in the Solanaceae and Rubiaceae; and, as their position in the Arabidopsis genome is known, they are useful tool for comparative mapping in diploid plant species (Wu et al. 2006). The augmented eggplant map comprises 289 orthologous markers (110 COSII + 179 tomato-derived RFLPs), spaced on average every 6.1 cM over a total map distance of 1535 cM (Wu et al. 2009a). Because of the syntenic nature of the tomato and eggplant genomes, the locations of an additional 522 COSII markers in the eggplant genome can be deduced. Thus, these researchers have calculated that a total of 869 molecular markers of known position are available in *S. melongena*.

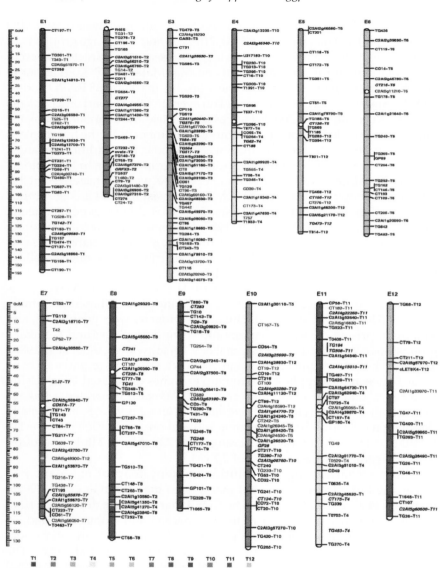

Figure 7-3 Genetic map of eggplant. Framework markers (LOD > 3) are in bold and by tick marks, interval markers (2 ≤ LOD ≤ 3) are in bold italics; all other markers are LOD < 2; cosegregating markers are alongside vertical bars. Chromosomal locations of markers on the tomato map are indicated by ~Tx after the marker name. Each tomato chromosome is color coded (see bottom of figure) and the corresponding segments of each eggplant linkage group are colored accordingly. (Used with permission from Wu et al. 2009a, Springer Science + Business Media.)

Color image of this figure appears in the color plate section at the end of the book.

While the molecular maps constructed thus far have employed segregating F_2 populations, the possibility of using a doubled haploid (DH) population for mapping AFLP markers in eggplant was explored by Barchi et al. (2010). F_1 hybrids derived from a cross between two eggplant breeding lines, one carrying a gene for *Fusarium* resistance introgressed from *S. aethiopicum* and the other a susceptible line were subjected to anther culture and 93 of the resulting DH individuals were screened with 170 AFLP markers. Of these, 68 percent showed distorted segregation making the population unusable for mapping analysis. In contrast, segregation ratios were not skewed in an F_2 population obtained by selfing the F_1 hybrids. Thus, 141 of these individuals were genotyped for 238 markers (212 AFLPs + 22 SSRs + 1 RFLP + 3 CAPS). The resulting map, constructed using JoinMap v. 4.0 (van Ooijen 2006), encompasses 718.7 cM on 12 linkage groups with an average density of one marker every 3.0 cM. The inclusion of 20 SSR markers from the Nunome et al. (2009) map permitted alignment and comparison of the maps; marker order and distances were found to be in close agreement.

10.2 Applications

10.2.1 Comparative Mapping

The close relationship of eggplant and potato to tomato, the first plant species for which a high density molecular linkage map was created (Tanksley et al. 1992), has facilitated comparative genomics within the Solanaceae. By using tomato RFLP markers as the basis for their eggplant map, Doganlar et al. (2002a) reported the existence of 28 chromosomal rearrangements, 23 inversions and five translocations in eggplant as compared to tomato. In contrast, only five inversions distinguish the tomato and potato genomes (Tanksley et al. 1992). Thus, eggplant is estimated to be three- to six-fold more diverged from tomato than potato; and the rate of divergence between eggplant and tomato approximates 0.19 rearrangements/chromosome/million years (assuming 12 million years of evolution between these two species (Wikstrom et al. 2001)). This is considered a rather moderate rate of evolution within angiosperms. Comparative mapping in tomato, potato and eggplant leads one to conclude that evolution in solanaceous genomes has largely involved paracentric inversions and translocations. Furthermore, fluorescence *in situ* hybridization (FISH) mapping of the entirety of chromosome 6 in seven *Solanum* species has given insight into the particulars of these changes; thus, one of the paracentric inversions previously thought to have occurred within the eggplant lineage is actually specific to the tomato genome (Lou et al. 2010).

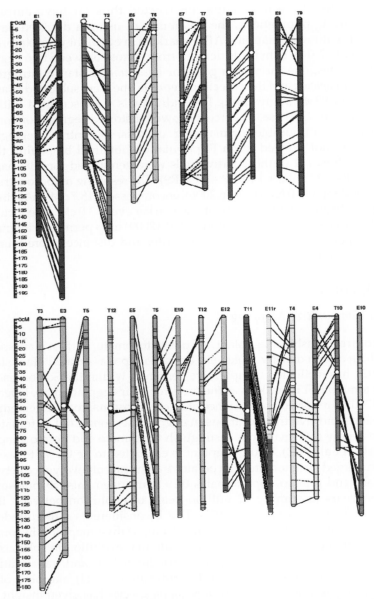

Figure 7-4 Comparative maps of the tomato and eggplant genomes. Color coding is the same as in Fig. 3. Centromere positions are indicated by white dots. Solid lines connect markers mapped at LOD ≥ 2 on both maps; dashed lines connect markers mapped at LOD < 2 on either map. (Used with permission from Wu et al. 2009a, Springer Science + Business Media.)

Color image of this figure appears in the color plate section at the end of the book.

The recent development of COSII markers in plants has paved the way for even broader comparisons among plant genomes (Wu et al. 2006). Anchored in the Arabidopsis genome, these markers show remarkable synteny (conservation of gene order and content), which spans 37 regions in the eggplant and tomato genomes (Wu et al. 2009a) (Fig. 7-4). Genetic analysis of the COSII markers has thus helped to explain the mechanisms of chromosome evolution in these two species. For example, the combined COSII-RFLP map reveals inversions on eggplant chromosomes 1, 2, 6, and 11 that were not detected in the RFLP map (Doganlar et al. 2002a). Furthermore, as these markers have also been mapped in pepper (Wu et al. 2009b), it can be inferred that while most of the inversions are specific to the eggplant lineage, some occurred during the divergence of potato and tomato from their most recent common ancestor with eggplant (Wu et al. 2009a). The comparative mapping efforts undertaken with the COSII markers provide important information about genome evolution as well as useful tools (i.e., molecular markers) for trait mapping and germplasm characterization.

10.2.2 Mapping and Tagging of Simply Inherited Traits

As a consequence of the relatively late development of genetic maps in eggplant, few genes have been positioned or tagged within its genome. For the most part, such studies have focused on genes controlling resistances to the major diseases and pests of eggplant. As levels of such resistance are typically low in cultivated materials within *S. melongena*, finding and tagging these resistances in wild germplasm has been a priority so that marker-assisted selection for these traits can be implemented in eggplant breeding programs.

Verticillium wilt resistance (derived from *S. linnaeanum*) was found linked to two genomic regions in the Sunseri et al. (2003) AFLP map. Bulked segregant analysis (BSA) of resistant and susceptible individuals was used to identify molecular markers linked to *Fusarium* resistance in eggplant (Mutlu et al. 2008; Toppino et al. 2008b). In one case, resistance genes were introduced into *S. melongena* lines from *S. aethiopicum* and *S. integrifolium* via somatic hybridization (Toppino et al. 2008b). RAPD markers linked to these resistance loci, which proved to be allelic, were then identified and used to create CAPS markers. Tests of the CAPS markers in segregating progenies indicated that they could help facilitate selection for the trait. Similarly, two sequence characterized amplified region (SCAR) markers were generated for a dominant *Fusarium* resistance gene introgressed into a Turkish eggplant line from a Malaysian genotype (Mutlu et al. 2008). The results of preliminary screens with these SCAR markers suggest that the gene might be specific to the Malaysian accession as it was not detected in the other *S. melongena* accessions and the wild species tested. BSA of F_2 progeny from a cross between Chinese inbred lines, one susceptible and one highly resistant to bacterial wilt (*Ralstonia solnancearum*), enabled tagging of the resistance gene (Cao et al. 2009). A RAPD fragment was then converted into a SCAR marker, which proved effective in distinguishing between resistant and susceptible plants. While generally considered a complex trait, peel color in eggplant was also tagged using bulked line analysis (BLA) (Liao et al. 2009). The development of a SCAR marker associated with reddish to dark purple pigmentation of the peel should facilitate selection for this key fruit quality trait. Advances in marker development and their application in eggplant will undoubtedly accelerate the rate at which economically important traits are tagged and increase the role of marker-assisted selection in breeding programs.

10.2.3 *Quantitative Trait Mapping*

Although tomato was a forerunner in the realm of quantitative trait mapping in plants, eggplant has lagged behind. Only as comprehensive linkage maps have become available has it been possible to study the inheritance of polygenic characters in this crop. The first such a trait mapped in eggplant was fruit shape which was analyzed in the *S. melongena* EPL1 × WCGR112-8 F_2 population developed by Nunome et al. (1998). A quantitative trait loci (QTL) for this character was detected on linkage group (LG) 10 of their RAPD map using the interval analysis function of Mapmaker/QTL v. 1.1 (Lincoln et al. 1992). In their combined RAPD-AFLP map, approximately 25 percent of the phenotypic variation in fruit shape was localized to the corresponding region of LG 2 (LG 10 having been re-named LG 2) (Nunome et al. 2001). The pigmentation of fruit, stem and calyx tissue was shown to be linked to loci on four linkage groups, with the most significant marker-trait association (explaining ~25, 50 and 68 percent of the phenotypic variation in fruit, stem and calyx color, respectively) on LG 7 (Nunome et al. 2001).

The genetic bases underlying a wide range of complex morphological traits were examined by Doganlar et al. (2002b) and Frary et al. (2003) in the *S. linnaeanum* MM195 x *S. melongena* MM738 F_2 mapping population. Fourteen traits related to domestication (fruit size and shape as well as the prickliness and color of leaves, stems and fruit) were measured in two locations (Doganlar et al. 2002b). A total of 62 significant ($P \leq 0.01$) QTLs were detected by linear point regression and confirmed by simple interval analysis (LOD ≥ 2.4) using QGene software (Nelson 1997). However, as growth conditions varied widely between the locations (greenhouse in New York vs. field in France), only 15 QTLs were detected as having a significant impact on corresponding traits at both locations. Nevertheless, it was possible to pinpoint the loci underlying traits of particular interest to eggplant breeders. Thus, two loci on linkage groups 2 and 9 controlled the majority of the variation in fruit size: *fw2.1* (23 percent PVE) and *fw9.1* (44 percent PVE). Similarly, fruit shape was determined by just two loci: *fs2.1* (20 percent PVE) and *fs7.1* (29 percent PVE). These same QTLs have been mapped in tomato (Grandillo et al. 1999; Frary et al. 2000) and pepper (Ben Chaim et al. 2001), indicating that these genes are conserved more broadly within Solanaceae. A single major locus on LG 6 determined the degree of plant prickliness (controlling as much as 79 percent of leaf prickle variation) while a single locus on LG 10 accounted for anthocyanin content in all plant tissues, explaining as much as 93 percent of the phenotypic variation. Daunay et al. (2004) suggest that this locus corresponds to their putative *A*, a dominant gene determining anthocyanin presence in the epidermis of various plant parts. Thus, the results of QTL analysis of domestication traits in eggplant support those in other crops; selection pressures exerted on a relatively small number of genetic loci over the course of domestication and cultivation have had profound and dramatic effects on plant form and function (Doebley et al. 1997).

In an analysis of 18 traits determining leaf, flower and fruit size, shape, appearance and development, Frary et al. (2003) identified 63 significant QTLs (P ≤ 0.01). Interestingly, at 46 percent of these loci, alleles from the wild parent were found to confer unexpected effects on phenotype. In several such cases, the wild alleles were responsible for producing transgressive effects, wherein the phenotype of the progeny was superior to that of either parent. For example, *S. linnaeanum* alleles were linked to improvements in flower and fruit number, fruit set, calyx size and fruit glossiness. These traits have major impacts on fruit yield and quality; thus, wild germplasm is a valuable genetic resource for the improvement of complex agronomic traits in cultivated eggplant. As with the domestication traits, some of the QTLs detected in eggplant correspond to loci controlling the quantitative inheritance of the same morphological characters in tomato and potato. This conservation of gene function and position has significant implications in terms of genome evolution within *Solanum*, and the subsequent discovery of extensive synteny between eggplant and tomato (Wu et al. 2009a) comes as less of a surprise in light of these earlier QTL-based studies. In practical terms, as tomato has been a model system for both classical and molecular genetics, much of the knowledge gained in tomato is readily transferrable to eggplant breeding and improvement.

11. Genomic Resources

Given eggplant's close relationship with the other solanaceous vegetable crops, researchers can easily leverage the extensive resources developed for tomato, potato and pepper and other nightshade family members for enhancing understanding of eggplant genetics and evolution. The SOL Genomics Network (SGN) (http://solgenomics.net/) is a relational database/website that curates all of the resources currently available for the Solananceae (Bombarely et al. 2011). SGN provides access to the molecular linkage maps of tomato, potato and eggplant as well as information about a range of marker types and links to mutant databases. Moreover, SGN provides all of the published gene or expressed sequence tag (EST) sequences from solanaceous species in a comparative genomics format as well as, at the time of writing, pre-publication draft assemblies of the genomes of *Solanum lycopersicum* (cultivated tomato) and *S. pimpinellifolium* (its wild progenitor). In addition to these resources, a set of 16,000 ESTs has been created from 60,000 cDNA clones from different developmental stages and tissues of eggplant (Fukuoka et al. 2010). While this unigene set is estimated to cover only ~35 percent of the transcriptome, it nevertheless provides a wealth of information specific to eggplant and with tremendous potential to impact work in comparative and functional genomics as well as phylogenetics in *Solanum*.

12. Future Prospects

There is much scope for future work in *Solanum melongena*. While still in its early stages, sequencing of eggplant's transcriptome will enable comparisons with other sequenced genomes thereby expanding resources for those interested in eggplant. An integrated molecular linkage map, which will combine RFLP, COSII and AFLP markers on the *S. linnaeanum* × *S. melongena* F$_2$ population is on the horizon (S. Doganlar, pers. comm.). With over 1,000 markers, this high density map should make it easier to explore aspects of gene and genome evolution in the Solanaceae as well as to locate the regions of the eggplant genome that control qualitative and quantitative traits of interest. Disease and pest resistances are much needed in the crop; thus, we can anticipate that marker-assisted selection will become an increasingly important tool in eggplant breeding as a means of facilitating the introgression of such genes from wild relatives. Breeding programmes are also likely to focus on improving the nutritional composition of the crop, particularly as regards anthocyanin and phenolic levels. Additional work to characterize antioxidant activity in cultivars, landraces and allied species will help in selecting the best germplasm for these traits. One can also imagine that the medicinal benefits of eggplant might lead to the plant being grown not as a vegetable crop but for its glycoalkaloids. As eggplant's nutritional benefits become more widely realized and higher yields are achieved through genetic improvement, eggplant should gain worldwide importance as a vegetable crop.

Acknowledgement

We are grateful to Anne Frary, Izmir Institute of Technology, Turkey, for valuable comments on the manuscript. This work was supported by a Career Project (TUBITAK 104T224) from the Scientific and Technical Research Council of Turkey to Sami Doganlar.

References

Acciarri N, Restaino F, Vitelli G, Perrone D, Zottini M, Pandolfini T, Spena A, Rotino GL (2002) Genetically modified parthenocarpic eggplants: improved fruit productivity under both greenhouse and open field cultivation. BMC plant Biol 2: 4.

Arpaia S, Mennella G, Onofaro V, Perri E, Sunseri F, Rotino GL (1997) Production of transgenic eggplant (*Solanum melongena* L.) resistant to Colorado potato beetle (*Leptinotarsa decemlineata* Say). Theor Appl Genet 95: 329–334.

Azuma K, Ohyama A, Ippoushi K, Ichiyanagi T, Takeuchi A, Saito T, Fukuoka H (2008) Structures and antioxidant activity of anthocyanins in many accessions of eggplant and its related species. J Agri Food Chem 56: 10154–10159.

Barchi L, Lanteri S, Portis E, Stàgel A, Valè G, Toppino L, Rotino GL (2010) Segregation distortion and linkage analysis in eggplant (*Solanum melongena* L.). Genome 53: 805–815.

Behera TK, Singh G (2002) Studies on resistance to shoot and fruit borer (*Leucinodes orbonalis*) and interspecific hybridization in eggplant. Indian J Hort 59: 62–66.

Behera TK, Sharma P, Singh BK, Kumar G, Kumar R, Mohapatra T, Singh NK (2006) Assessment of genetic diversity and species relationships in eggplant (*Solanum melongena* L.) using STMS markers. Sci Hort 107: 352–357.

Ben Chaim A, Paran I, Grube RC, Jahn M, van Wijk R, Peleman J (2001) QTL mapping of fruit-related traits in pepper (*Capsicum annuum*). Theor Appl Genet 102: 1016–1028.

Bennett MD, Leitch IJ (2010) Plant DNA C-values database (release 7.0, Dec. 2010) http://www.kew.org/cvalues/, 1 September 2011.

Bitter G (1923) Solana Africana, Part IV, Repert Spec Nov Regni Veg Beih 16: 1–320.

Bletsos FA, Roupakias DG, Thanassoulopoulos CC (2000) Gene transfer from wild *Solanum* species to eggplant cultivars: prospects and limitations. In: van der Plas LHW [ed] XXV International Horticultural Congress, Part 12: Applications of Biotechnology and Molecular Biology and Breeding—General Breeding, Breeding and Evaluation of Temperate Zone Fruits for the Tropics and the Subtropics. Acta Hort 522: 71–78.

Bletsos F, Thanassoulopoulos C, Roupakias D (2003) Effect of grafting on growth, yield and *Verticillium* wilt of eggplant. HortScience 38: 183–186.

Bletsos FA, Stavropoulos NI, Papdopoulou PD (2004) Evaluation of eggplant (*Solanum melongena* L.) germplasmfor resistance to *Verticillium* wilt. Adv Hort Sci 18: 33–37.

Bombarely A, Menda N, Tecle IY, Buels RM, Strickler S, Fischer-York T, Pujar A, Leto J, Gosselin J, Mueller LA (2011) The Sol Genomics Network (solgenomics.net): growing tomatoes using Perl. Nucl Acids Res 39: D1149–1155.

Borgato L, Conicella C, Pisani F, Furini A (2007) Production and characterization of arboreous and fertile *Solanum melongena* + *Solanum marginatum* somatic hybrid plants. Planta 226: 961–969.

Broun P, Tanksley SD (1996) Characterization and genetic mappingof simple repeat sequences in the tomato genome. Mol Gen Genet 250: 39–49.

Cao BH, Lei JJ, Wang Y, Chen GJ (2009) Inheritance and identification of SCAR marker linked to bacterial wilt-resistance in eggplant. Afr J Biotechnol 8: 5201–5207.

Cham BE (2007) Solasodine rhamnosyl glycosides specifically bind cancer cell receptors and induce apoptosis and necrosis treatment for skin cancer and hope for internal cancers. Res J Biol Sci 2: 503–514.

Collonnier C, Fock I, Kashyap V, Rotino GL, Daunay MC, Lian Y, Mariska IK, Rajam MV, Servaes A, Ducreux G, Sihachakr D (2001) Applications of biotechnology in eggplant. Plant Cell Tiss Org Cult 65: 91–107.

Collonnier C, Fock I, Mariska I, Servaes A, Vedel F, Siljak-Yakovlev S, Souvannavong V, Sihachakr D (2003) GISH confirmation of somatic hybrids between *Solanum melongena* and *S. torvum*: assessment of resistance to both fungal and bacterial wilts. Plant Physiol Biochem 41: 459–470.

Daunay MC (2008) Eggplant. In: Prohens J, Nuez F [eds] Handbook of Crop Breeding, Vegetables II: Fabaceae, Liliaceae, Umbelliferae, and Solanaceae. Springer, New York, USA pp 163–220.

Daunay MC, Janick J (2007) History and iconography of eggplant. Chron Hort 47: 16–22.

Daunay MC, Maggioni L, Lipman E (2003) [compilers] Solanaceae Genetic Resources in Europe. Report of two meetings -21 September 2001, Nijmegen, Netherlands and 22 May 2003, Skierniewice, Poland. International Plant Genetic Resources Institute, Rome, Italy.

Daunay MC, Lester RN, Ano G (2001) Eggplant. In: Charrier A, Jacquot M, Hamon S, Nicolas D [eds] Tropical Plant Breeding. Science Publishers, Montpellier, France pp 199–222.

Daunay MC, Aubert S, Frary A, Doganlar S, Lester RN, Barendse G, van der Weerden G, Hennart JW, Haanstra J, Dauphin F, Jullian E (2004) Eggplant (*Solanum melongena*) fruit colour: pigments, measurements and genetics. In: Proceedings of the 7th EUCARPIA Meeting on Genetics and Breeding of Capsicumand Eggplant,. Noordwijkerhout, Netherlands pp 108–116.

Daunay MC, Laterrot H, Janick J (2007) Iconography of the Solanaceae from antiquity to the XVIIth century: a rich source of information on genetic diversity and uses. In: Spooner DM, LM Bohs, J Giovannoni, RG Olmstead, D Shibata ed VIth International Solanaceae Conference: Genomics Meets Biodiversity. Acta Hort 745: 59–88.

Doebley J, Stec A, Hubbard L (1997) The evolution of apical dominance in maize. Nature 386: 485–488.

Doganlar S, Frary A, Daunay MC, Lester RN, Tanksley SD (2002a) A comparative genetic linkage map of eggplant (*Solanum melongena*) and its implications for genome evolution in the Solanaceae. Genetics 161: 1697–1711.

Doganlar S, Frary A, Daunay MC, Lester RN, Tanksley SD (2002b) Conservation of gene function in the Solanaceae as revealed by comparative mapping of domestication traits in eggplant. Genetics 161: 1713–1726.

Donzella G, Spena A, Rotino GL (2000) Transgenic parthenocarpic eggplants: superior germplasmfor increased winter production. Mol Breed 6: 79–86.

FAO Statistics (2009) http://faostat.fao.org (Cited 2009), 1 September 2011.

Frary A, Nesbitt TC, Frary A, Grandillo S, van der Knapp E, Cong B, Liu J, Meller J, Elber R, Alpert KA, Tanksley SD (2000) *fw2.2*: a quantitative trait locus key to the evolution of tomato fruit size. Science 289: 85–88.

Frary A, Doganlar S, Daunay MC, Tanksley SD (2003) QTL analysis of morphological traits in eggplant and implications for conservation of gene function during evolution of Solanaceous species. Theor Appl Genet 107: 359–370.

Frary A, Doganlar S, Daunay MC (2007) Eggplant. In: Kole C [ed] Genome Mapping and Molecular Breeding in Plants, Vol 5: Vegetables Springer-Verlag, Berlin pp 231–257.

Fukuoka H, Yamaguchi H, Nunome T, Negoro S, Miyatake K, Ohyama A (2010) Accumulation, functional annotation, and comparative analysis of expressed sequence tags in eggplant (*Solanum melongena* L.), the third pole of the genus *Solanum* species after tomato and potato. Gene 450: 76–84.

Furini A, Wunder J (2004) Analysis of eggplant (*Solanum melongena*)-related germplasm morphological and AFLP data contribute to phylogenetic interpretations and germplasm utilization. Theor Appl Genet 108: 197–208.

Gangopadhyay KK, Mahajan RK, Kumar G, Yadav SK, Meena BL, Pandey C, Bisht IS, Mishra SK, Sivaraj N, Gambhir R, Sharma SK, Dhillon BS (2010) Development of a core set in brinjal (*Solanum melongena* L.). Crop Sci 50: 755–762.

Gebhardt SE, Thomas RG (2002) Nutritive value of foods. Home Garden Bulletin (USDA) 72: 80–81.

Gisbert C, Prohens J, Raigón MD, Stommel JR, Nuez F (2011) Eggplant relatives as sources of variation for developing new rootstocks: effects of grafting on eggplant yield and fruit apparent quality and composition. Sci Hort 128: 14–22.

Goggin FL, Jia LL, Shah S, Hebert S, Williamson VM, Ullman DE (2006) Heterologous expression of the *Mi-1.2* gene from tomato confers resistance against nematodes but not aphids in eggplant. Mol Plant Microbe Interact 19: 383–388.

Gousset C, Collonnier C, Mulya K, Mariska I, Rotino GL, Besse P, Servaes A, Sihachakr D (2005) *Solanum torvum*, as a useful source of resistance against bacterial and fungal diseases from improvement of eggplant (*S. melongena* L.). Plant Sci 168: 319–327.

Grandillo S, Ku HM, Tanksley SD (1999) Identifying the loci responsible for natural variation in fruit size and shape in tomato. Theor Appl Genet 99: 978–987.

Hall CA, Hobby T, Cipollini M (2006) Efficacy and mechanisms of solasonine- and solamargine-induced cytolysis on two strains of *Trypanosoma cruzii*. J Chem Ecol 32: 2405–2416.

Hamilton GC, Jelenkovic GL, Lashomb JH, Ghidiu G, Billings S, Patt JM (1997) Effectiveness of transgenic eggplant (*Solanum melongena* L.) against the Colorado potato beetle. Adv Hort Sci 11: 189–192.

Hanson PM, Yang RY, Tsou SCS, Ledesma D, Engle L, Lee TC (2006) Diversity in eggplant (*Solanum melongena*) for superoxide scavenging activity, total phenolics, and ascorbic acid. J Food Compos Anal 19: 594–600.

Isshiki S, Kawajiri N (2002) Effect of the cytoplasm of *Solanum violaceum* Ort. on fertility of eggplant (*Solanum melongena* L.). Sci Hort 93: 9–18.

Isshiki S, Taura T (2003) Fertility restoration of hybrids between *Solanum melongena* L. and *S. aethiopicum* L. Gilo Group by chromosome doubling and cytoplasmic effect on pollen fertility. Euphytica 134: 195–201.

Isshiki S, Okubo H, Fujieda K (2000) Segregation of isozymes in selfed progenies of a synthetic amphidiploid between *Solanum integrifolium* and *S. melongena*. Euphytica 112: 9–14.

Isshiki S, Iwata N, Khan MMR (2008) ISSR variations in eggplant (*Solanum melongena* L.) and related *Solanum* species. Sci Hort 117: 186–190.

Iwamoto Y, Hirai M, Ohmido N, Fukui K, Ezura H (2007) Fertile somatic hybrids between *S. integrifolium* and *S. sanitwongsei* (syn. *S. kurzii*) as candidates for bacterial wilt resistant rootstock of eggplant. Plant Biotechnol 24: 179–184.

Janick J, Topoleski LD (1963) Inheritance of fruit color in eggplant (*Solanum melongena*). Proc Amer Soc Hort Sci 83: 547–558.

Jarl CI, Rietveld EM, de Haas JM (1999) Transfer of fungal tolerance through interspecific somatic hybrisation between *Solanum melongena* and *S. torvum*. Plant Cell Rep 18: 791–796.

Kameya T, Miyazawa N, Toki S (1990) Production of somatic hybrids between *Solanum melongena* L. and *S. integrifolium* Poir. Jpn J Breed 40: 429–434.

Karihaloo JL, Gottlieb LD (1995) Allozyme variation in the eggplant, *Solanum melongena* L. (Solanaceae). Theor Appl Genet 90: 578–583.

Karihaloo JL, Brauner S, Gottlieb LD (1995) Random amplified polymorphic DNA variation in the eggplant, *Solanum melongena* L. (Solanaceae). Theor Appl Genet 90: 767–770.

Kaur C, Kapoor HC (2001) Antioxidants in fruits and vegetables—the millennium's health. Int J Food Sci Technol 36: 703–725.

Kakizaki Y (1931) Hybrid vigor in eggplants and its practical utilization. Genetics 16: 1–25.

Khah EM (2005) Effect of grafting on growth, performance and yield of aubergine (*Solanum melongena* L.) in the field and greenhouse. J Food Agri Environ 3: 92–94.

Khan MMR, Isshiki S (2008) Development of a male sterile eggplant by utilizing the cytoplasm of *Solanum virginianum* and a biparental transmission of chloroplast DNA in backcrossing. Sci Hort 117: 316–320.

Khan MMR, Isshiki S (2010) Development of a cytoplasmic male-sterile line of eggplant (*Solanum melongena* L.) with the cytoplasm of *Solanum anguivi*. Plant Breed 130: 256–260.

Kikuchi K, Honda I, Matsuo S, Fukuda M, Saito T (2008a) Stability of fruit set in newly selected parthenocarpic eggplant lines. Sci Hort 115: 111–116.

Kikuchi K, Honda I, Matsuo S, Fukuda M, Saito T (2008b) Evaluation of fruit set and fruit development of new Japanese parthenocarpic eggplant lines. In: Cohen J, Lee JM [eds] XXVII International Horticultural Congress—IHC 2006: International Symposium on Endogenous and Exogenous Plant Bioregulators. Acta Hort 774: 121–126.

Kumar PA, Mandaokar K, Sreenivasu SK, Chakrabati S, Bisari S, Sharma SR, Kaur S, Sharma RP (1998) Insect-resistant transgenic brinjal plants. Mol Breed 4: 33–37.

Kuo KW, Hsu SH, Li YP, Lin WL, Liu LF, Chang LC, Lin CC, Lin CN, Sheu HM (2000) Anticancer activity evaluation of the *Solanum* glycoalkaloid solamargine: triggering apoptosis in human hematoma cells. Biochem Pharmacol 60: 1865–1873.

Lander ES, Green P, Abrahamson J, Barlow A, Daly MJ, Lincoln SE, Newberg LA (1987) MAPMAKER: an interactive computer package for constructing primary genetic linkage maps of experimental and natural populations. Genomics 1: 174–181.

Lester RN, Hasan SMZ (1991) Origin and domestication of the brinjal eggplant, *Solanum melongena* from *S. incanum*, in Africa and Asia. In: Hawkes JG, Lester RN, Nee M, Estrada RN ed Solananceae III. Taxonomy, Chemistry, Evolution. Royal Botanic Garden, Kew, UK pp 369–387.

Lester RN, Jaeger PML, Bleijendaal-Spierings BHM, Bleijendaal HPO, Holloway HLO (1990) African eggplants: a review of collecting in West Africa. Plant Genet Resour Newsl. 81/82: 17–26.

Levin RA, Myers NR, Bohs L (2006) Phylogenetic relationships among the "spiny solanums" (*Solanum* subgenus *Leptostemonum*, Solanaceae). Amer J Bot 93: 157–169.

Li H, Chen H, Zhuang T, Chen J (2010) Analysis of genetic variation in eggplant and related *Solanum* species using sequence-related amplified polymorphism markers. Sci Hort 125: 19–24.

Liao Y, Sun BJ, Sun GW, Liu HC, Li ZL, Wang GP, Chen RY (2009) AFLP and SCAR markers associate with peel color in eggplant (*Solanum melongena*). Agri Sci China 8: 1466–1474.

Lincoln S, Daly M, Lander E (1992) Mapping genes controlling quantitative traits with MAPMAKER/QTL. Whitehead Institute Technical Report, 2nd edn, Cambridge, Massachusetts, USA.

Liu RH (2003) Health benefits of fruit and vegetables are from additive and synergistic combinations of phytochemicals. Amer J Clin Nutr 78: 517S–520S.

Lou Q, Iovene M, Spooner DM, Buell CR, Jiang J (2010) Evolution of chromosome 6 of *Solanum* species revealed by comparative fluorescence *in situ* hybridization mapping. Chromosoma 119: 435–442.

Mace ES, Lesterand RN, Gebhardt CG (1999) AFLP analysis of genetic relationships among the cultivated eggplant, *Solanum melongena* L., and wild relatives (Solanaceae). Theor Appl Genet 99: 626–633.

Mao W, Jinxin Y, Sihachakr D (2008) Development of core subset for the collection of Chinese cultivated eggplants using morphological-based passport data. Plant Genet Resour Characteriz Utiliz 6: 33–40.

Mennella G, Rotino GL, Fibiani M, D'Alessandro A, Francese G, Toppino L, Cavallanti F, Acciarri N, Lo Scalzo R (2010) Characterization of health-related compounds in eggplant (*Solanum melongena* L.) lines derived from introgression of allied species. J Agri Food Chem 58: 7597–7603.

Muñoz-Falcón JE, Prohen J, Vilanova S, Nuez F (2008) Characterization, diversity, and relationships of the Spanish striped (*Listada*) eggplants: a model for the enhancement and protection of local heirlooms. Euphytica 164: 405–419.

Mutlu N, Boyaci FH, Göçmen M, Abak K (2008) Development of SRAP, SRAP-RGA, RAPD and SCAR markers linked with a *Fusarium* wilt resistance gene in eggplant. Theor Appl Genet 117: 1303–1312.

Nelson JC (1997) QGENE: software for marker-based genomic analysis and breeding. Mol Breed 3: 229–235.

Nisha P, Abdul Nazar P, Jayamurthy P (2009) A comparative study on antioxidant activities of different varieties of *Solanum melongena*. Food Chem Toxicol 47: 2640–2644.

Noda Y, Kaneyuki T, Igarashi K, Mori A, Packer L (2000) Antioxidant activity of nasunin, an anthocyanin in eggplant. Toxicology 148: 119–123.

Nolla JAB (1932) Inheritance of color in eggplant. Journal of the Department of Agriculture, Puerto Rico 16: 19–30.

Nunome T, Yoshida T, Hirai M (1998) Genetic linkage map of eggplant. In: Proceedings of the 10th Eucarpia Meeting on Genetics and Breeding of *Capsicum*and Eggplant, Avignon, France pp 239–242.

Nunome T, Ishiguro K, Yoshida T, Hirai M (2001) Mapping of fruit shape and color development traits in eggplant (*Solanum melongena* L.) based on RAPD and AFLP marker. Breed Sci 51: 19–26.

Nunome T, Suwabe K, Ohyama A, Fukuoka H (2003a) Characterization of trinucleotide microsatellites in eggplant. Breed Sci 53: 77–83.

Nunome T, Suwabe K, Iketani H, Hirai M (2003b) Identification and characterization of microsatellites in eggplant. Plant Breed 122: 256–262.

Nunome T, Negoro S, Kono I, Kanamori H, Miyatake K, Yamaguchi H, Ohyama A, Fukuoka H (2009) Development of SSR markers derived from SSR-enriched genomic library of eggplant (*Solanum melongena* L.). Theor Appl Genet 119: 1143–1153.

Okmen B, Sigva HO, Mutlu S, Doganlar S, Yemenicioglu A, Frary A (2009) Total antioxidantactivity and total phenolic contents in different Turkish eggplant (*Solanum melongena* L.) cultivars. Int J Food Prop 12: 616–624.

Pal JK, M Singh, M Rai, S Satpathy, DV Singh, S Kumar 2009 Development and bioassay of *Cry1Ac*-transgenic eggplant (*Solanum melongena* L.) resistant to shoot and fruit borer. J Hort Sci Biotechnol 84: 434–438.

Polignano G, Uggenti P, Bisignano V, Della Gatta C (2010) Genetic divergence analysis in eggplant (*Solanum melongena* L.) and allied species. Genet Resour Crop Evol 57: 171–181.

Prabhavati V, Yadav JS, Kumar PA, Rajam MV (2002) Abiotic stress tolerance in transgenic eggplant (*Solanum melongena* L.) by introduction of bacterial mannitol phosphodehydrogenase gene. Mol Breed 9: 137–147.

Prohens J, Blanca JM, Nuez F (2005) Morphological and molecular variation in a collection of eggplants from a secondary center of diversity: implications for conservation and breeding. J Amer Soc Hort Sci 130: 54–63.

Prohens J, Rodríguez-Burruezo A, Raigón MD, Nuez F (2007) Total phenolic concentration and browning susceptibility in a collection of different varietal types and hybrids of eggplant: implications for breeding for higher nutritional quality and reduced browning. J Am Soc Hortic Sci 132: 638–646.

Prohens J, Muñoz-Falcón A, Rodríguez-Burruezo A, Nuez F (2008) Strategies for the breeding of eggplants with improved nutritional quality. In: Prange RK, SD Bishop [ed] XXVII International Horticultural Congress—IHC2006: International Symposium of Sustainability through Integrated and Organic Horticulture. Acta Horticulturae 767 pp 285–291.

Raigón MD, Prohens J, Muñoz-Falcón JE, Nuez F (2008) Comparison of eggplant landracesand commercial varieties for fruit content of phenolics, minerals, dry matter and protein. J Food Compos Anal 21: 370–376.

Rajam MV, Kumar SV (2007) Eggplant. In: Pua EC, Davey MR [eds] Biotechnology in Agriculture and Forestry. Vol 59: Transgenic Crops IV, Springer, New York, USA pp 201–219.

Ribeiro AP de O, Pereira EJG, Galvan TL, Picanco MC, Picoli EA de T, da Silva DJH, Fari MG, Otoni WC (2006) Effect of eggplant transformed with oryzacystatin gene on *Myzus persicae* and *Macrosiphum euphorbiae*. J Appl Entomol 130: 84–90.

Rizza F, Mennella G, Collonnier C, Sihachakr D, Kashyap V, Rajam MV, Presterà M, Rotino GL (2002) Androgenic dihaploids from somatic hybrids between *Solanum melongena* and

S. aethiopicum L. group *gilo* as a source of resistance to *Fusarium oxysporum* f. sp. *melongenae*. Plant Cell Rep 20: 1022–1032.

Rotino GL, Perri E, Zottini M, Sommer H, Spena A (1997) Genetic engineering of parthenocarpic plants. Nat Biotechnol 15: 1398–1401.

Rotino GL, Sihachakr D, Rizza F, Valè G, Tacconi MG, Alberti P, Mennella G, Sabatini E, Toppino L, D'Alessandro A, Acciarri N (2005) Current status in production and utilization of dihaploids from somatic hybrids between eggplant (*Solanum melongena* L.) and its wild relatives. Acta Physiol Plant 27: 723–733.

Sadder MT, Al-Shareef RM, Hamdan H (2007) Assessment of genetic, morphological and agronomical diversity among Jordanian eggplant (*Solanum melongena* L.) landraces using random amplified polymorphic DNA (RAPD). In: Spooner DM, Bohs LM, Giovannoni J, Olmstead RG, Shibata D [eds] VIth International Solanaceae Conference: Genomics Meets Biodiversity, Acta Horticulturae. 745 pp 303–310.

Sadilova E, Stintzing FC, Carle R (2006) Anthocyanins, colour and antioxidant properties of eggplant (*Solanum melongena* L.) and violet pepper (*Capsicum annuum* L.) peel extracts. Z Naturforsch C 61: 537–535.

Saito T, Matsunaga H, Saito A, Hamato N, Koga T, Suzuki T, Yoshida T (2009a) A novel source of cytoplasmic male sterility and a fertility restoration gene in eggplant (*Solanum melongena* L.) lines. J Jpn Soc Hort Sci 78: 425–430.

Saito T, Yoshida T, Monma S, Matsunaga H, Sato T, Saito A, Yamada T (2009b) Development of the parthenocarpic eggplant cultivar 'Anominori'. Jpn Agri Res Quart 43:123–127

Sakata Y, Lester RN (1997) Chloroplast DNA diversity in brinjal eggplant (*Solanum melongena* L.) and related species. Euphytica 97: 295–301.

Sakata Y, Monma S, Narikawa T, Komochi S (1996) Evaluation of resistance to bacterial wilt and *Verticillium* wilt in eggplants (*Solanum melongena* L.) collected in Malaysia. J Jpn Soc Hort Sci 65: 81–88.

Sánchez-Mata M-C, Yokoyama WE, Hong Y-J, Prohens J (2010) α–solasonine and α–solamargine contents of gboma (*Solanum macrocarpon* L.) and scarlet (*Solanum aethiopicum* L.) eggplants. J Agri Food Chem 58: 5502–5508.

Schaff DA, Jelenkovic G, Boyer CD, Pollack BL (1982) Hybridization and fertility of hybrid derivatives of *Solanum melongena* L. and *Solanum macrocarpon* L. Theor Appl Genet 62: 149–153.

Shelton AM (2010) The long road to commercialization of Bt brinjal (eggplant) in India. Crop Protec 29: 412–414.

Shiu LY, Chang LC, Liang CH, Huang YS, Sheu HM, Kuo KW (2007) Solamargine induces apoptosis and sensitizes breast cancer cells to cisplatin. Food Chem Toxicol 45: 2155–2164.

Singh AK, Singh M, Singh AK, Singh R, Kumar S, Kalloo G (2006) Genetic diversity within the genus *Solanum* (Solanaceae) as revealed by RAPD markers. Curr Sci 90: 711–716.

Stàgel A, Portis E, Toppino L, Rotino GL, Lanteri S (2008) Gene-based microsatellite development for mapping and phylogeny studies in eggplant. BMC Genomics 9: 357.

Stam P (1993) Construction of integrated genetic linkage map by means of a new computer package: JOINMAP. Plant J 3: 739–744.

Stommel JR, Whitaker BD (2003) Phenolic acid content and composition of eggplant fruit in a germplasm core subset. J Amer Soc Hort Sci 128: 704–710.

Sunseri F, Sciancalepore A, Martelli G, Acciarri N, Rotino GL, Valentino D, Tamietti G (2003) Development of RAPD-AFLP map of eggplant and improvement of tolerance to *Verticillium* wilt. In: Hammerschlag FA, Saxena P [eds] XXVI International Horticultural Congress: Biotechnology in Horticultural Crop Improvement. Acta Horticulturae 625 pp 107–115.

Swarup V (1995) Genetic resources and breeding of aubergine (*Solanum melongena* L.). Acta Hort 412: 71–79.

Tanksley SD, Ganal MW, Prince JP, de Vicente MC, Bonierbale MW, Broun P, Fulton TM, Giovannoni JJ, Grandillo S, Martin GB, Messeguer R, Miller JC, Miller L, Paterson AH,

Pineda O, Roder S, Wing RA, Wu W, Young ND (1992) High-density molecular linkage maps of the tomato and potato genomes. Genetics 132: 1141–1160.

Tigchelaar EC, Janick J, Erickson HT (1968) The genetics of anthocyanin coloration in eggplant (*Solanum melongena* L.). Genetics 60: 475–491.

Tiwari SK, Karihaloo JL, Hameed N, Gaikwad AB (2009) Molecular characterization of brinjal (*Solanum melongena* L.) cultivars using RAPD and ISSR markers. J Plant Biochem Biotechnol 18: 1–7.

Toppino L, Mennella G, Rizza F, D'Alessandro A, Sihachakr, Rotino GL (2008a) ISSR and isozyme characterization of androgenetic dihaploids reveals tetrasomic inheritance in tetraploid somatic hybrids between *Solanum melongena* and *Solanum aethiopicum* group *Gilo*. J Hered 99: 304–315.

Toppino L, Valè G, Rotino GL (2008b) Inheritance of *Fusarium* wilt resistance introgressed from *Solanum aethiopicum* Gilo and *Aculeatum* groups into cultivated eggplant (*S. melongena*) and development of associated PCR-based markers. Mol Breed 22: 237–250.

Tümbilen Y, Frary A, Daunay MC, Doganlar S (2011a) Application of EST SSRs to examine genetic diversity in eggplant and its close relatives. Turk J Biol 35: 125–136.

Tümbilen Y, Frary A, Mutlu S, Doganlar S (2011b) Genetic diversity in Turkish eggplant (*Solanum melongena*) varieties as determined by morphological and molecular analyses. Int Res J Biotechnol 2: 16–25.

van Ooijen J (2006) JoinMap® 4, software for the calculation of genetic linkage maps in experimental populations. Kyazma BV, Wageningen, The Netherlands.

Wang J-X, Gao T-G, Knapp S (2008) Ancient Chinese literature reveals pathways of eggplant domestication. Ann Bot 102: 891–897.

Weese TL, L Bohs (2010) Eggplant origins: out of Africa, into the Orient. Taxon 59: 49–56.

Wikstrom N, Savolainen V, Chase MW (2001) Evolution of the angiosperms: calibrating the family tree. Proc Roy Soc Lond, Ser B: Biol Sci 268: 2211–2220.

Wivutvongvana M, Lumyong P, Wivutvongvana P (1984) Exploration and collection of eggplant germplasm in northern and northeastern Thailand. IBPGR Newsl 8: 6–10.

Wu FN, Mueller LA, Crouzillat D, Petiard V, Tanksley SD (2006) Combining bioinformatics and phylogenetics to identify large sets of single-copy orthologous genes (COSII) for comparative, evolutionary and systematic studies: a test case in the euasterid plant clade. Genetics 174: 1407–1420.

Wu FN, Eannetta NT, Xu Y, Tanksley SD (2009a) A detailed synteny map of the eggplant genome based on conserved ortholog set II (COSII) markers. Theor Appl Genet 118: 927–935.

Wu FN, Eannetta NT, Xu YM, Durrett R, Mazourek M, Jahn MM, Tanksley SD (2009b) A COSII map of the pepper genome provides a detailed picture of synteny with tomato and new insights into recent evolution in the genus *Capsicum*. Theor Appl Genet 118: 1279–1293.

Index

Color Plate Section

Chapter 1

Figure 1-3 Species identification on the basis of morphology. The flower traits are the most distinct between species. All *C. annuum* and *C. chinense* have a white corolla. *C. frutescens* has blue anthers and the corolla color varies from white to greenish-white. The corolla color of *C. baccatum* is white with yellow spots. In *C. pubescens*, both corolla and stamen colors are purple. *C. chacoense* has white and small corollas. Additionally, the seed color of *C. pubescens* accessions is black, while for most *Capsicum* species seed is tan.

Chapter 2

Figure 2-1 Genetic diversity showing various fruit shapes and colors in pepper germplasm.

Figure 2-2 Morphology of anthers in fertile (left) and CMS (right) and pepper lines. Figures were quote from Min (2009).

Chapter 3

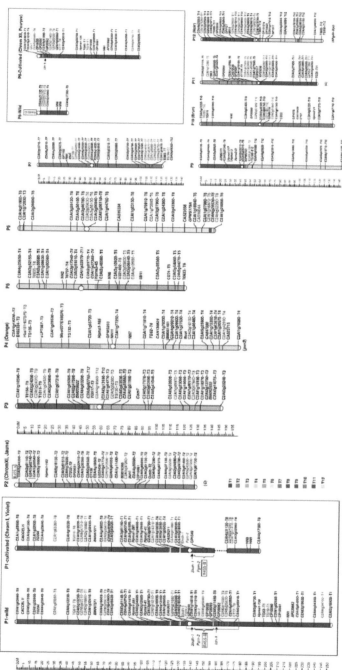

Figure 3-1 A genetic map of pepper based on the COSII markers. The pepper linkage groups are designated P1–P12 based on the synteny with tomato chromosomes and are based on the chromosome names from the trisomic analysis of Pochard (1977). Each tomato chromosome is assigned with one color according to the color code below P2 and the corresponding pepper chromosomes are shown with the same color. ~TN following pepper markers indicates their corresponding tomato chromosomes. The putative centromeric regions are indicated by the white dot. More details on this map can be found in Wu et al. (2009). [Reprinted from Wu et al. (2009) by permission.]

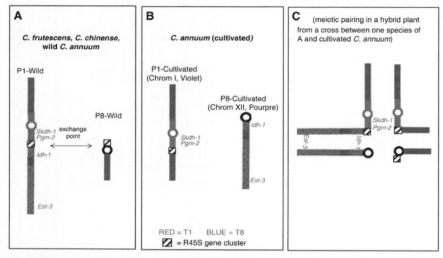

Figure 3-2 A model describing the translocation between the cultivated *C. annuum* and the wild *C. annuum* and the related *Capsicum* species *C. chinense* and *C. frutescens*. [Reprinted from Wu et al. (2009) by permission.]

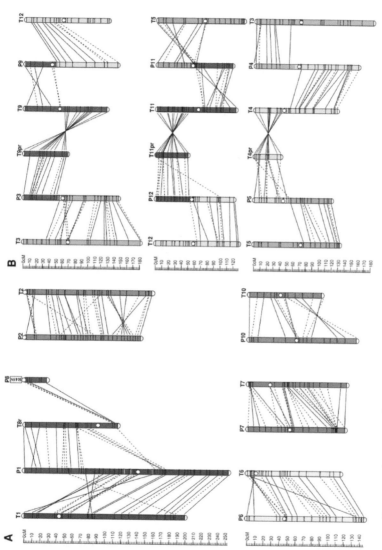

Figure 3-3 A comparative map of pepper and tomato. The chromosome colors are coded as in Fig. 1. The orthologous markers are connected by lines. More details on this map can be found in Wu et al. (2009). [Reprinted from Wu et al. (2009) by permission.]

Chapter 5

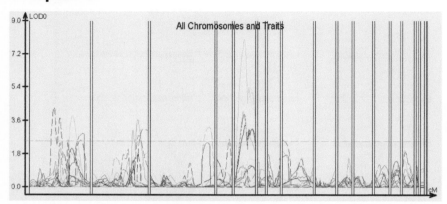

Figure 5-9 Whole-genome scan of pepper by QTL Cartographer for QTLs controlling 12 horticulturally important traits in pepper (Jorgensen C and Prince JP, unpublished results). Notice the LOD score significance threshold below 3.0.

Chapter 6

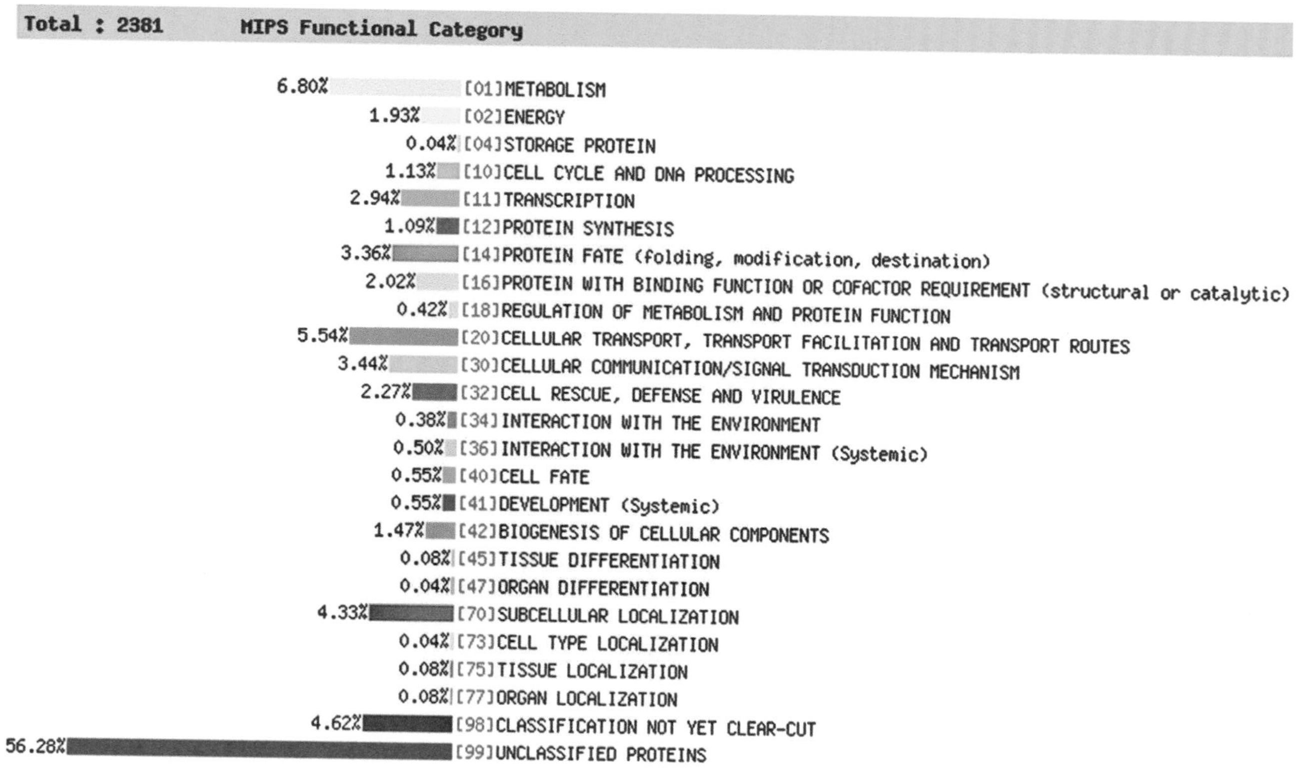

```
Total : 2381        MIPS Functional Category

6.80%    [01]METABOLISM
    1.93%    [02]ENERGY
        0.04% [04]STORAGE PROTEIN
        1.13% [10]CELL CYCLE AND DNA PROCESSING
        2.94% [11]TRANSCRIPTION
        1.09% [12]PROTEIN SYNTHESIS
        3.36% [14]PROTEIN FATE (folding, modification, destination)
        2.02% [16]PROTEIN WITH BINDING FUNCTION OR COFACTOR REQUIREMENT (structural or catalytic)
        0.42% [18]REGULATION OF METABOLISM AND PROTEIN FUNCTION
    5.54%    [20]CELLULAR TRANSPORT, TRANSPORT FACILITATION AND TRANSPORT ROUTES
    3.44%    [30]CELLULAR COMMUNICATION/SIGNAL TRANSDUCTION MECHANISM
    2.27%    [32]CELL RESCUE, DEFENSE AND VIRULENCE
        0.38% [34]INTERACTION WITH THE ENVIRONMENT
        0.50% [36]INTERACTION WITH THE ENVIRONMENT (Systemic)
        0.55% [40]CELL FATE
        0.55% [41]DEVELOPMENT (Systemic)
        1.47% [42]BIOGENESIS OF CELLULAR COMPONENTS
        0.08% [45]TISSUE DIFFERENTIATION
        0.04% [47]ORGAN DIFFERENTIATION
    4.33%    [70]SUBCELLULAR LOCALIZATION
        0.04% [73]CELL TYPE LOCALIZATION
        0.08% [75]TISSUE LOCALIZATION
        0.08% [77]ORGAN LOCALIZATION
    4.62%    [98]CLASSIFICATION NOT YET CLEAR-CUT
56.28%       [99]UNCLASSIFIED PROTEINS
```

Figure 6-1 MIPS functional categories of the pepper ESTs.

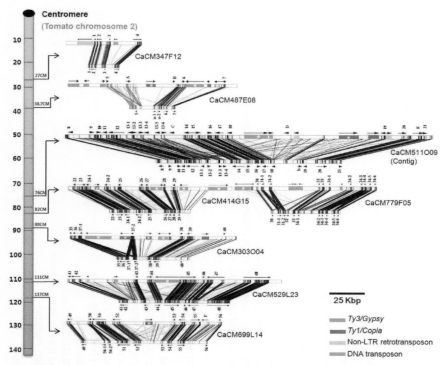

Figure 6-5 Sequence comparisons between orthologous gene-rich regions of pepper and tomato. The arrows indicate the predicted genes. The accession numbers indicate the orthologous gene sets. Colored bars indicate the repeat elements. The vertical lines indicate the highly similar regions.

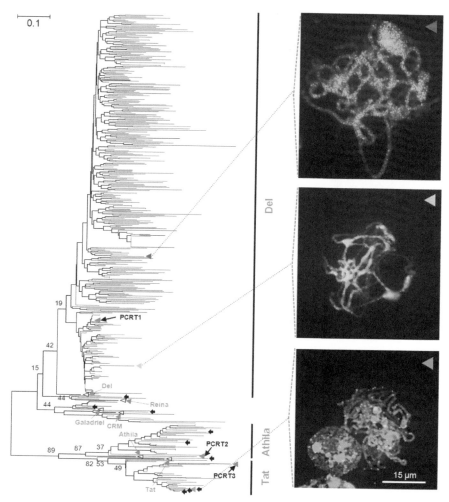

Figure 6-7 Phylogenetic analysis of pepper and tomato *Ty3/Gypsy*-like elements. The pepper and tomato *Ty3/Gypsy*-like elements are depicted by red and blue lines, respectively. The major subgroups Tat, Athila and Del are indicated by green letters. The reverse transcriptases (RTs) used as FISH probes are marked with triangles (purple, yellow and green). The FISH result for each of the probes is indicated by a dotted line. The black arrows indicate the RTs found in the selected gene-rich pepper sequences. The empty black triangles indicate the representative RTs for the different subgroups acquired from the GyDB.

Chapter 7

Figure 7-2 The exterior and interior of fruit of two eggplant hybrids: a transgenic parthenocarpic line containing the *iaaM* gene (P10) and a control (C10). (Used with permission from Acciarri et al. 2002, licensee BioMed Central Ltd.)

Figure 7-3 Genetic map of eggplant. Framework markers (LOD > 3) are in bold and by tick marks, interval markers (2 ≤ LOD ≤ 3) are in bold italics; all other markers are LOD < 2; cosegregating markers are alongside vertical bars. Chromosomal locations of markers on the tomato map are indicated by ~Tx after the marker name. Each tomato chromosome is color coded (see bottom of figure) and the corresponding segments of each eggplant linkage group are colored accordingly. (Used with permission from Wu et al. 2009a, Springer Science + Business Media.)

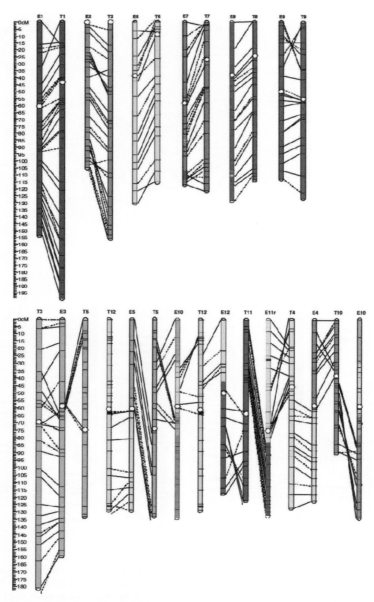

Figure 7-4 Comparative maps of the tomato and eggplant genomes. Color coding is the same as in Fig. 3. Centromere positions are indicated by white dots. Solid lines connect markers mapped at LOD ≥ 2 on both maps; dashed lines connect markers mapped at LOD < 2 on either map. (Used with permission from Wu et al. 2009a, Springer Science + Business Media.)

*For Product Safety Concerns and Information please contact
our EU representative GPSR@taylorandfrancis.com Taylor & Francis
Verlag GmbH, Kaufingerstraße 24, 80331 München, Germany*

T - #0031 - 160425 - C12 - 234/156/10 [12] - CB - 9781466577459 - Gloss Lamination